GARDEN PESTS
& PREDATORS

Explains how you can encourage beneficial organisms and discourage harmful pests within a healthy garden community. Includes clear illustrations to enable you to distinguish friend from foe.

Photography for cover by Geoff du Feu

Pictures show: Lacewing (*top*); Ladybird Larva Eating Blackfly (*bottom left*); Small White Butterfly (*bottom right*).

GARDEN PESTS & PREDATORS

How to Identify Them and How to Encourage the Good and Discourage the Bad

by

Elfrida Savigear

Illustrated by Isobel Elsey

THORSONS PUBLISHERS LIMITED

Wellingborough, Northamptonshire

First published 1980

© ELFRIDA SAVIGEAR 1980

British Library Cataloguing in Publication Data

Savigear, Elfrida
 Garden pests and predators.
 1. Garden pests
 2. Pest control—Biological control
 I. Title
 635'.04'996   SB603.5

 ISBN 0-7225-0554-X
 ISBN 0-7225-0553-1 Pbk

Photoset by Specialised Offset Services Limited, Liverpool
and printed in Great Britain by
Weatherby Woolnough, Wellingborough, Northamptonshire

CONTENTS

PREFACE

Gardens are living places. There is no need to go to a zoo or nature reserve to see a thriving array of wildlife – your garden is filled with millions of organisms which live in harmony most of the time, but as in any population, there are the mis-fits.

You are also part of this living ecosystem, and it is likely that you are the one organism with the greatest influence, because you choose to grow certain plants, remove some vegetation to eat elsewhere and may add manure or fertilizers from another area. This creates an artificial situation.

No doubt you have seen gardens which are left to run wild, and have observed that the plants include those weeds you spend many hours removing. So the organisms which depend upon that land for their food and homes will be different too – and if the land is continually left to itself the organisms will balance out until a stable state results.

But what about your garden, which you want to be attractive, highly productive and easy to maintain? How can you encourage beneficial organisms, discourage harmful pests and grow what you please? This book tries to help you become aware of the living creatures so close at hand, showing you how you can encourage a healthy garden community.

1

GARDEN RESIDENTS

In the ordinary garden there are two distinct habitats, the soil and the air. (You may have a glasshouse which produces a modified environment, and reference will be made to glasshouse organisms in the general text). These two environments have distinct differences which affect the type of organism found.

Characteristics of Soil Organisms
Soil organisms may like the dark, often stuffy atmosphere. Quite often they will die if exposed to the light or air which is drier and richer in oxygen.

Many of the smallest soil organisms (the microscopic ones) move by swimming, and need the soil moisture to get from one place to another. Several soil organisms are anaerobic which means they are able to respire without oxygen. The soil environment has more gradual temperature changes than the air especially deeper in the soil, so during the summer months it is a cooler place to live, and during the winter it can be the place for avoiding frosts. Mainly for this reason many eggs and larva develop underground, but it is also possible for these juvenile stages to be sheltered from their enemies by remaining under stones and rotting vegetation. Thus many soil organisms are primitive with few protective features.

Characteristics of Aerial Organisms
The organisms which live above the ground display many features more varied than those confined or living mainly underground.

Environmental Factor	Direct Effect	Result for Organisms
LIGHT	Photosynthesis (Food making by green plants)	Plant growth greatest in the Summer which will encourage an increase in the organisms feeding on plant material
	Dormancy	Plants and animals may be dormant over Winter
	Flowering and Fruiting	Organisms such as bees and butterflies attracted to the flowers and fruit.
OXYGEN CONTENT	Respiration (Converting food into energy)	Aerial organisms can be far more active than soil organisms.
LOW RELATIVE HUMIDITY	Most living organisms are at least 80% water	Require some method to control internal water supplies e.g. hard exoskeleton of beetles.
TEMPERATURE VARIATIONS	Extremes can cause death	Require a temperature control system e.g. sweat glands of mammals or organisms have the ability to move to cooler, sheltered areas.

The obvious differences in the above-ground environment are the diurnal and seasonal light changes, a much lower concentration of carbon dioxide with a higher concentration of oxygen, a low relative humidity (except when it rains!), large temperature variations, air movement ranging to gale force winds, and a far larger variety of moving organisms to interact with each other. This variety means it is difficult to make generalizations. It is easy to observe that plants are green and use the light to photosynthesize, but can any features of other organisms be linked to the aerial environment?

Biological classification is a complicated procedure and it is normal to specialise on one particular facet such as entomology (the study of insects). But the garden contains organisms from almost every group of living things, so classification and identification may seem a daunting prospect. A gardener needs to be able to place an organism within a general group, and from there identify how it lives and interacts with the rest of the garden.

This general classification is simplified to direct you to the main groups — you should quickly master the characteristics, and then go straight to the chapter concerned.

Large Organisms: Vertebrates
These are probably only visitors to most small gardens, but many town gardens provide homes for rodents, hedgehogs and various birds. The presence of a backbone (vertebrae) and an internal skeleton produces organisms with less protective covering so they can usually move fast and can carry their food away to a more protected environment. Most of these organisms are well-known and will be considered in the following groups related to their community activity.

1. Dog, Cat.
2. Fox, Badger, Mole, Rabbit.
3. Squirrel, Rats, Mice.
4. Hedgehog.
5. Toad, Frog.
6. Tits, Thrush, Blackbird, Seagull, Starling.
7. Pigeon, Bullfinch, Chaffinch, Heron, Sparrow.

These macro-organisms are important because their size enables even a single example to have a large influence in a garden, whether due to trampling or earthworks as in the case of burrowing vertebrates, or due to the voracious appetites which need to be satisfied. Animals in general can be classified as follows:

> Carnivorous i.e. feed on animal material.
> Herbivorous i.e. feed on vegetation.
> Omnivorous i.e. feed on animal or vegetable material.

The carnivores are generally beneficial, whereas the herbivores on a vegetarian diet may cause much damage to plant material.

Worms Etc.

The ordinary *earthworm* needs no introduction to the gardener, and although there are different species typical of different soil types, they generally influence the garden environment in a similar way. Most gardeners are pleased to see worms, although the exceptionally 'lawn proud' may dislike the casts, and feel worms should be discouraged.

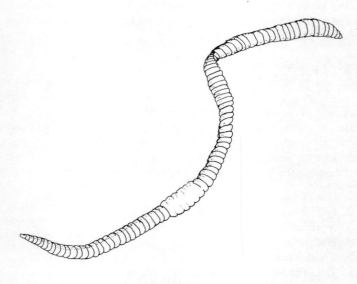

Earthworm.

There are various other creatures which have been included in the wormtype classification. These are the *potworms* which are smaller than earthworms and usually grey or white in colour, and are commonly found in compost heaps.

Potworm.

Nematodes are non-segmented worms which are only about 1mm in length and almost colourless, so in normal circumstances are considered microscopic. Gardeners will come across nematodes (also known as eelworms) which are pests, but usually only the damage provides evidence of the eelworm visitor. There are other nematodes which feed on bacteria and help in the general interaction of communities.

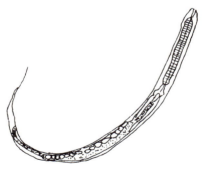

Nematode.

Millepedes and Centipedes are long and thin like worms, and they have segmented bodies. They also have many legs (more than fifteen pairs), the millepedes with two pairs per segment, the centipede with one pair per segment.

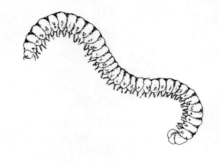

Millepede.

Centipede.

Symphyla These are similar to centipedes but they are whitish in colour and only have twelve pairs of legs.

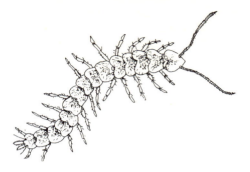

Symphylid.

Woodlice have been included in this group because they have more legs than the insect group and curl up in a similar fashion to millepedes. They have seven pairs of thoracic legs, with flattened limbs extending from the abdomen.

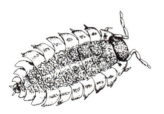

Woodlouse.

Insect larvae often appear wormlike and may be found in the sheltered soil environment. They can be distinguished from the worm group in that although they are segmented, they have less than fifteen segments. They will be considered in the insect section (see Chapter 6).

Slugs

The mollusc group, or slugs and snails, are probably familiar to most people, although individual species are less likely to be well known. The distinctive characteristics are a muscular foot used for movement along a slime trail, and the presence of a shell. The shell is exterior in the snail and interior within the slug.

Spiders

The large hairy spiders which occupy the bath are, in fact, a great asset to the house as their diet consists of the less hygienic visitors like blow flies. Similarly, garden spiders should be your friends. They can be distinguished from other organisms, especially the insects, by having two main body parts, and four pairs of jointed legs. The very small spider-like organisms or mites include some plant pests.

Spider.

Insects

This group of organisms contains many thousands of different species and there is a tremendous amount of variety making identification a job for a specialist. Many insects have more than one appearance during their development, and some may be pests at one stage and beneficial at another. However the most common garden insects have been sub-divided into major types which are fairly easily recognised, with a final mixed group containing important extras.

The majority of creepy-crawlies belong to the insect group, but to be sure of correct classification, they should at some stage in their life cycle show the features shown on the diagram.

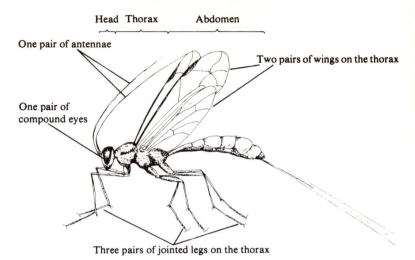

Head Thorax   Abdomen

One pair of antennae

Two pairs of wings on the thorax

One pair of
compound eyes

Three pairs of jointed legs on the thorax

Typical Adult Insect.

Aphid Group (e.g. Greenfly, Blackfly). Aphids can usually be identified because they reproduce very rapidly to produce colonies which cover soft, fleshy plant material as they suck out the sugar from the plant. They may often be wingless.

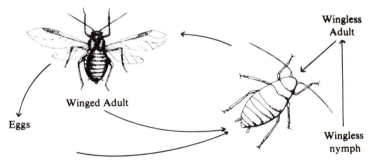

Wingless
Adult

Winged Adult

Eggs

Wingless
nymph

Aphid Life Cycle (Incomplete Metamorphosis).

The nymph, which looks very similar to the adult although it is smaller, will never have wings and is unable to reproduce. This type of life cycle is known as *incomplete metamorphosis*, and it means identification is easier as the insect has a similar appearance throughout all its life.

Butterfly Group. This group is the first example demonstrating *complete metamorphosis*, which means that the insect is found to be completely different in looks and life-style during the different stages of its development. This can be represented by the following life cycle diagram:

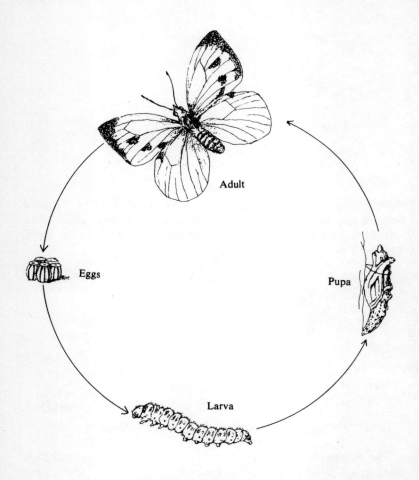

Butterfly Life Cycle (Complete Metamorphosis).

Fly Group. The true flies have most of the typical insect features (as the adult butterfly) but they only have one pair of wings and vestigial (withered) hind wings known as haltares. The life cycle is represented in diagrammatic form, the larvae which are frequently pests are fairly primitive, as they tend to live in a fairly sheltered situation.

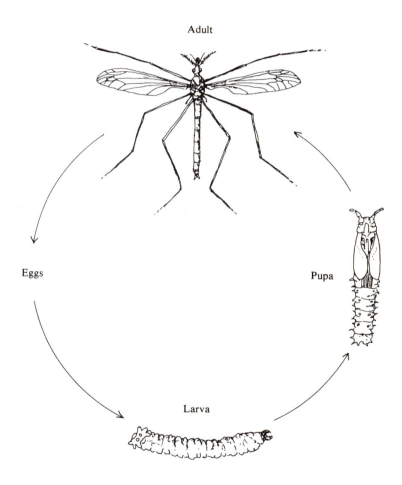

Fly Life Cycle (Complete Metamorphosis).

Beetle Group. The Beetles are the insects in tough armour as you may have noticed when you have tried to squash one. The armour is produced by a toughening of the front wings to produce *elytra*, which usually covers the folded hind wings and abdomen. The mouth parts show well developed mandibles.

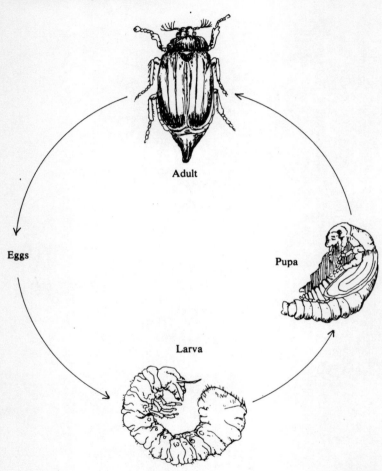

Beetle Life Cycle (Complete Metamorphosis).

Bee Group. This fairly diverse group includes the bees, wasps, sawflies and ants, so both the friend and foe of the gardener. The adults show normal insect structure but the two pairs of wings are clear and membranous, and there is a distinct 'waist' between the thorax and abdomen.

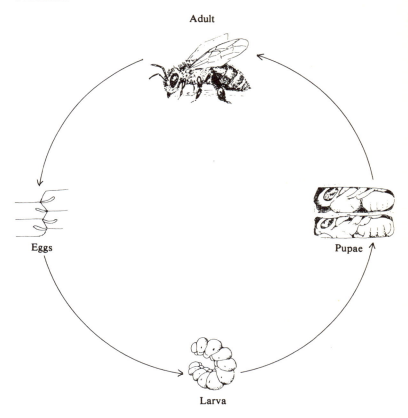

Bee Life Cycle (Complete Metamorphosis).

Other Insects. Some insects do not fit into any of the previous groups, but must be included because they have a distinct influence on the garden environment. Four examples are described in this section, the lacewings, dragonflies, earwigs and bush crickets.

Microscopic Organisms

The largest number of organisms in the garden, especially in the soil, will be included in the microscopic section. Evidence of their activity is easily found, such as the tremendous heat produced within a heap of grass cuttings. But as they cannot be seen they can easily be ignored, whereas, as with all living things, they have their likes and dislikes. It is not normally necessary to identify these micro-organisms to the finest detail, but it is important to understand their life styles so that, wherever possible, the beneficial ones are encouraged and the harmful ones discouraged. The types of organisms considered more closely in Chapter 7 are:

Microscopic algae
Bacteria
Protozoa
Fungi
Mycorrhiza
Viruses

The fungi may produce macroscopic structures such as toadstools, but the feeding parts which can be harmful or beneficial, are microscopic.

2

VERTEBRATES

Most of the animals included in this section are visitors rather than permanent residents, although this will depend upon the particular individual and where the garden is situated.

Household Pets

Household pets which enter the garden may be of varied description — I daresay at least one of my readers possesses snakes or an elephant or ...? But the most common domestic animals to enter the garden must be dogs and cats. Are they good or bad? Like many of the other organisms we will mention, they are both beneficial and harmful. Both animals are naturally clean and will make use of the garden for excretory purposes. Bitches tend to use the same area of grass on many occasions, and this will eventually kill the vegetation due to the high concentration of nutrients which prevent the plants taking up water. Dogs tend to go 'little and often' which, unless it happens to be a territory marking such as a gateway hedge, usually produces less conspicuous results.

Tom cats are very similar to dogs in their behaviour, but cats in general try to bury their excrement and any freshly dug soil (especially where seeds have been planted!) is a great attraction. They usually seem to use every garden except their own, which is rather unfortunate for the non cat-lover. Cats also like to catch birds and sometimes they will steal fledgelings and prevent them growing to adulthood.

The main benefit of these animals is no doubt their companionship, but they also serve a useful function in warding off pests. Some birds delight to eat freshly planted seeds or young seedlings, others are partial to fresh raspberries and plums, and here the dog or cat will probably need no encouragement to chase them off. Both animals are useful in catching rodents such as rats or mice, although we humans are not always too pleased when the catch is presented inside the house.

Burrowing Vertebrates

The next few vertebrate animals are considered together because their main influence upon the garden is similar due to their burrowing activities. Most of them are mainly found in the larger gardens and more rural areas, but foxes especially are becoming much more common in town gardens. From the gardener's point of view they must all be classed as pests, but some naturalists would be quite pleased to have a close 'viewing' site.

Foxes. Foxes live in earths, which are semi-underground dens preferably in sandy soil, but they will also take over another animal's home such as a large rabbit hole. Cubs are usually born in Spring and the dog will remain with the vixen during her pregnancy and for a while after the cubs are born. Foxes feed on rats, mice, rabbits, birds, frogs, earthworms and beetles, but more recently they have become scavengers at dustbins and refuse tips, and they have always been known to take the occasional lamb.

Badgers. Badgers are really only visitors in the rural areas and as they are very shy creatures, few people see them. They live in sets, which are underground homes in well drained soil usually in a wooded situation. They feed during the night on a mixed diet including rabbits, earthworms, slugs and snails, fruit, nuts and grass. They are useful predators, and in most garden situations they need not be destroyed.

Moles. Moles are also shy creatures, but their presence is soon announced by molehills, the small mounds of earth produced while they dig their underground tunnels. They have specialized 'spade' forelimbs which easily move the soil out of the way. While digging, the mole will disturb earthworms, insects, millepedes and slugs and snails, and these make up its diet.

Rabbits. Rabbits are the last main burrowing garden visitors, and once more are confined to rural or semi-rural areas. Their warrens can be large especially in a sandy bank, but they are unlikely to burrow in the more formal type of garden. They are herbivorous, and may be a nuisance grazing new shoots and even removing the bark from young trees.

Rodents

The rodents are a large group of mammals most of whom are shy but they may leave a visiting card of some sort.

Grey Squirrels. Grey squirrels are very common visitors to gardens if there are any large deciduous trees around where they can make their nest or drey. They are especially fond of acorns, but their diet is wide-stretching including other types of nut, cones and other fruits, bulbs, grain, plant material and bark, fungi and various insects.

Rats. Rats always present a picture of unpleasant events such as plagues, and most criticisms have some justification. Few gardens will be visited by rats except where there are neglected buildings providing shelter. They traditionally feed on grain, but their diet includes insects, mice and plant material, as well as scavenging around waste places.

Mice. Mice are much more common visitors and although they create fear in many humans, they are really fairly harmless. Wood mice live in a tunnel system amongst rotting vegetation so they can be disturbed in the compost heap. They are active during the dark hours, when they search out seeds as their most popular food. At other seasons of the year they may feed on insect larvae.

The house mouse as the name suggests likes a more sheltered existence, but they may sometimes be found in garden sheds. They are omnivorous, enjoying plant material, insects and flesh from larger mammals too.

Voles. Voles can also be quite common in the garden during some years, making their nests in long vegetation or amongst straw or other mulches at the base of plants. This gives the timid creatures shelter, and they will gnaw away at the base of trees and shrubs. If this continues all the way round a tree it will damage the growing tissue or cambium and successfully kill the plant. They are able to feed on tough plant material because they are equipped with the normal chisel-like biting teeth, but also a very efficient set of grinders. Some species are more fond of other plant material such as berries, buds and bulbs, but fortunately they are themselves popular prey to the household cat!

Hedgehogs

Hedgehogs are very distinctive creatures and most people have a soft spot for them. Many of them become quite tame, and call regularly for a supper of bread and milk. Gardeners are wise to encourage hedgehogs as they feed on many of our problem pests, but perhaps it should be pointed out that they harbour many fleas, and dogs and cats will pick these up although they are not so fond of the domestic host. Their hair has become modified into sharp spines which protect them from all predators (except the motor car).

Hedgehog – a friend to be encouraged.

If a hedgehog falls onto its back the spines will not hurt it as each spine is embedded in a thick layer of muscular material. They nest in sheltered shrubbery both when they are caring for their young, mainly during the summer, and also when they hibernate during the winter months. Their diet is varied but they are particularly partial to slugs, earthworms, insects, mice, frogs, berries and other plant material. They are noisy foragers so they may often be heard grunting and snorting in the undergrowth.

Toads and Frogs

Toads. Toads and frogs are easily recognized but quite often people find it hard to distinguish the two. Toads are more solidly built than frogs, with shorter hind legs and a dull dry pimply skin enabling them to merge into the background. They will rest camouflaged for many hours and this enables them to survive the attacks of many would-be predators. They jump using all four feet and this makes them appear rather clumsy. They produce an offensive fluid when attacked by birds, and this leaves them safer from enemy attack.

Toad — with a voracious appetite.

Toads feed on many different fauna including beetles, caterpillars, snails, worms, woodlice and mice, and seem to have a continual appetite. They establish a definite 'home' under a stone or sheltered by a root, and although they may travel some distance to feed, they usually return to their homes. They need a pond for their gelatinous string of spawn, but the remainder of their life is on land and mainly controlled by the abundance of insect life as a food source.

Frogs. Frogs have more smooth yellowish green skins and long hind legs adapted for jumping. Once more water is essential for breeding and tadpole development, and damp muddy places are required for winter hibernation. But frogs are voracious feeders on slugs, worms and insects, and they can jump their way some distance from the nearest pond. Obviously you are most likely to see frogs if you have your own pond but frogs from a neighbour's pond will quite likely visit your garden and should certainly be encouraged.

Frog, hungry for slugs, worms and insects.

Beneficial Birds

Most people are delighted to see birds in a garden and British people spend a lot of money on various foods to encourage bird visitors. Many birds are beneficial too as they are predacious on insect pests, and even when soft fruit is all eaten by our feathered friends, they seem to be forgiven in a way that slugs or caterpillars never are.

Blackbirds are one of the commonest garden birds and they are especially welcome because of their tuneful songs. They are very partial to earthworms which could possibly be beneficial if they remove them from lawns, but is mainly detrimental to soil structure. They also eat many larvae and adult insects and so seem more welcome than not.

Blackbird: a helpful songster.

Starlings are not such popular visitors as they tend to be noisy and untidy and often frighten off smaller, more attractive birds. They are very fond of brightly coloured berries, but it is possible to protect plants by nets, and their diet also includes insects, especially leatherjackets.

Starling: noisy but devouring many insect larvae.

Thrushes are welcome visitors, especially if your garden is troubled by snails. Quite commonly a tapping sound in the garden will be a thrush banging a snail against a stone, having a juicy morsal as its reward. Thrushes do not restrict their diet to snails and they eat many other insects and small animals, as well as enjoying fruit as a sweetmeat.

Thrush: fond of snails.

Robins seem a very British institution and because they have distinct territories, they may become quite tame. They like any soft juicy morsel which means they will not stop to decide whether their prey is beneficial or harmful! However, as one of their favourite foods are caterpillars we can perhaps forgive them for eating the occasional worm.

Tits are delightful birds partly because of their colourful attire, but also because they are somewhat cheeky. Their diet is almost entirely confined to insects so they are high on the list of beneficial birds. If the winter is hard, most insects or larvae will be destroyed or well hidden, so tits will be grateful for a supply of nuts and also a bowl of unfrozen water! Do not worry if tits appear to be pecking at unopened buds — they are gently looking for insects and insect eggs and rarely cause any harm.

Seagulls are also garden visitors, and they use their beaks to hook out insects and larvae which are sheltering beneath the soil.

Feathered Pests

Unfortunately some birds must really be classed as garden pests because they do such a lot of damage. Top of the list may be the bullfinch, one of our most colourful birds, but using beauty as a mask for devilry! They have fairly soft beaks which means they need to feed on softer material. They are often to be found enjoying soft fruit, and, more especially, attacking the blossoms and buds of many trees and shrubs. Sometimes they work so quickly pecking out the buds of cherry trees, that it is hard to believe they have time to enjoy anything.

Bullfinch: a colourful devil.

Chaffinches may also damage flower buds but this is usually less common. Sparrows are often a nuisance in spring time when they are particularly attracted to crocuses. This is for a good reason, as the flowers contain a supply of vitamin A, but it is annoying to have one of our earliest spring flowers pulled to pieces.

Pigeons seem always to make their presence known by their powerful coos, a sound which may seem pleasant at first but can become very irritating after a while. They can be a dreadful nuisance by searching out newly planted seeds and also feeding on seeds such as peas and beans. During the winter months they will peck away at winter greens, and of course, they frighten off smaller birds and can empty a bird table very rapidly!

Pigeon: hungry for seeds.

Herons are beautiful birds and I am always thrilled to watch one flying over. They are, however, becoming common garden pests or thieves, stealing goldfish from ponds.

Heron: a visitor looking for goldfish.

3

WORM GROUP

This chapter tells of the activities of garden organisms which are fairly small, are long and thin in shape and tend to wriggle or crawl around the ground. The most common of these is probably the earthworm which has a very interesting life history.

The Earthworm

Earthworms have been observed by many naturalists and Darwin wrote a book about worm activity on plant material. Scientists are still busy investigating worm action, and the more work proceeds, the more important and intriguing worms prove to be.

There are just over twenty species of earthworm found in Britain. Worms characteristically have up to one hundred and fifty cylindrical segments, and four pairs of bristles on the lower half of almost every segment. They are pinkish brown in colour and usually have a distinct band or *clitellum* about one third of the way from the head end. The skin possesses glands which produce mucus over the surface of the worm to prevent desiccation and to facilitate movement through the soil. There are light receptive cells in the worm's skin, and these warn

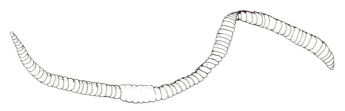

Earthworm, showing saddle or clitellum.

it to burrow down to safety again when you bring it to the surface by
your digging. There is a simple brain at the mouth end of the worm
and a main nerve which passes all the way down its body.

Reproduction

Earthworms are hermaphrodite possessing both male and female sex
organs in different parts of their bodies. But they still have to mate to
exchange their sperm with another worm. The commonest earthworm
mates above ground, but the majority of species hide away
underground. The two worms become bound together with mucus, the
sperm are exchanged and later the worms separate. The clitellum
produces a skin-like secretion which gradually hardens and the worm
then wriggles backwards out of this tube which closes to become the
cocoon.

The eggs are in the cocoon first, and the sperm picked up as the
cocoon moves over the worm's body. The eggs are fertilised
externally, and the young worm develops to emerge one to five
months later. The young worms tend to be colourless on hatching but
usually show all their segments. They continue to grow at rates
varying with temperature and moisture, and will be sexually mature in
one to two years. They may live for six or so years, but this is very
dependent upon predators, disease and chemicals.

Regeneration

It is a common view that if you cut a worm in half, two worms will be
produced. Regeneration can occur but on many occasions you are
more likely to kill the worm. It occurs most readily when a worm loses
a few of the segments at its posterior end, but it will not occur if too
much of the tissue is damaged. New growth is also most efficient in
warm temperatures and when worms are young.

The Gardener's Friend

Earthworms are beneficial organisms to the normal soil because they
burrow through the soil which increases the aeration and drainage,
and they also feed on, and break down, organic material such as
leaves. They burrow through the soil by alternately contracting and
relaxing the longitudinal and circular muscles in the body wall. As
they move, in fact, they eat the soil in front of them and this is mixed
with moisture and other waste food material and excreted as the cast.
Within the burrow, the cast is pressed against the burrow walls, and

some fungi will grow on this releasing nutrients and breaking down the organic matter even further.

The shape and depth of the burrow depends on the species of earth worm but most of them are active in the first 300mm (12 inches) of the soil. If the weather is very cold or very dry, they will burrow deeper to find a more amenable clime. They feed on plant and fungal material and the preference seems to be for soft deciduous leaves if they are available, but some species are fond of animal dung. All worms need calcium carbonate as they secrete this with their digestive juices, and this is probably the reason why they are normally absent from acid soils and do not like eating leaves from coniferous trees.

You can encourage earthworms in various ways. They are most active in fairly well drained, alkaline loam soils where there is plenty of organic matter for them to feed upon. They will like your activities in the vegetable plot and their burrows will bring air to the roots encouraging good growth. They continue to mix the soil up and bring nutrient-rich soil up to the surface. They are very active beneath grass and here they help produce a stone-free crumb soil.

It is unfortunate that worm-casts on lawns cause some gardeners to kill worms because their lawn will become less well drained and they will have to spike the lawn instead of allowing the worms to do the work for them. One fairly effective way to discourage worms from the surface is to treat your lawn with ammonium sulphate fertiliser. This supplies nitrogen which will encourage green healthy growth, but it also makes the soil more acid which will make the worms stay below ground.

Potworms

Potworms are most commonly seen as active components of the compost heap. They are small grey or whitish worms which wriggle actively when they are disturbed. They tend to be found in groups around good food sources, and they seem to enjoy rotting plant material, fungi and bacteria and eelworms. They are thought to control some eelworm pests, by feeding on young stages.

Potworm: tiny, white worms.

Nematodes

Nematodes or roundworms are found in an amazing variety of habitats, but the majority continue life without us noticing them. It is within this group that many of our 'worm' parasites are found, and this is partly because they feed by absorbing liquid food, in this case from their host. Others are found in the garden mainly as harmless soil organisms, but sometimes feeding on plants and causing damage. Most of these nematodes are very tiny and almost colourless, so they certainly do not draw attention to their presence.

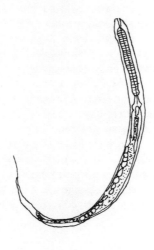

Nematode or eelworm.

The Nematode Life Cycle

The life cycle is fairly simple and straight forward. Male and female worms mate and eggs are laid protected by a tough protein wall. The eggs hatch to produce tiny worms protected by a tough cuticle. As the juvenile worm grows, it has to moult because the cuticle is too tough to expand. It moults four times before it is adult. Nematodes move by an eel-like swimming motion brought about by contraction and relaxation of the muscular body against the tough protective cuticle.

Feeding Habits

The nematodes which live freely in the soil (and some estimations have found more than ten million per square metre of soil) feed mainly in the top 50mm (2 ins) and feed on algae, protozoa and bacteria and rotting organic matter and other nematodes. They do not have any obvious effect upon the soil, but as they feed on other organisms they must be important in maintaining a balanced society.

Parasitic Nematodes

There are several different types of nematodes which are considered pests, but the ones which are probably the most widespread are known as migratory root eelworms. These are parasitic on plant roots at some stage during their lives, but they can also swim from plant to plant and then they are possibly the second largest offenders for virus transmission (aphids top of the list!) They differ from normal soil nematodes by having a hollow spear inside their mouths, which they use to pierce plant cells before they inject their digestive juices and suck out the sap.

Nematodes or eelworms feeding on roots are very difficult to spot and frequently other organisms such as fungi or bacteria enter where the nematode has damaged the plant tissue. This continues to make the plant stunted and look sick, but if the plant is pulled up the rotted roots fall off and the eelworms remain in the soil to swim off to some new site of attack. The best way of control is obviously to avoid the pest, but soil sterilants do give fairly effective control. Alternatively it is possible to grow *Tagetes* species and plough these into the soil before a susceptible crop is planted – this reduces the population of eelworms.

Cyst Eelworms

These eelworms are important pests because they are so difficult to control. They enter the root at the second larval stage and they swim between the cells until they reach the centre of the root where the cells of the root are rich in nutrients. The eelworms produce saliva which makes the plant cells greatly enlarge and results in a weak and stunted plant.

Leaf Eelworms

Leaf eelworms feed on leaf tissue by entering through the leaf stomatal pores. They swim from one area to another making use of dew, rain or water from the can or hose. One of the commonest is Chrysanthemum eelworm, and the first visible signs are when yellow discolourations appear between the veins of the lower leaves. Later on the leaves will turn black and flop down against the stem. Any eelworms within the dead tissue, dry up and become curled like a watch-spring, and can then remain dormant for several months. When moisture is available they can continue as before.

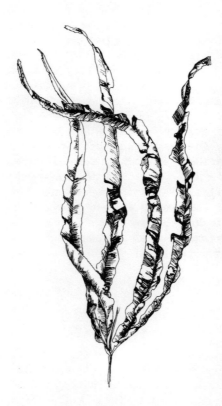

Eelworm damage to fern fronds.

Stem Eelworms

Stem eelworms enter the plant through small wounds or through plant pores such as lenticels. Once within the plant they feed on the plant tissues and cause swellings and commonly rotting.

Millepedes

Millepedes are related to insects as they have jointed legs but of course, they have many more legs, (even though a 'thousand' is rather exaggerated). Their bodies are constructed of between twenty and sixty segments most of which have two pairs of legs.

There are two main groups of harmful millepedes commonly referred to as the snake millepedes and the flat millepedes. The snake millepedes have hard cylindrical bodies and they live in the soil feeding on organic matter, and plant roots, (especially where the roots have already been weakened by some other organism) or soft material such as seedlings or germinating seeds.

The flat millepedes are shorter with flatter bodies and longer legs – these are the ones which are often found in the greenhouse. Pill millepedes are also common having received a descriptive name as they roll into a ball when they are disturbed. Millepedes tend to be far less active than centipedes, so if a many legged worm-like creature disappeared from view, it was likely to be a centipede!

Millepedes have poorly developed mouthparts which means they are unable to feed on any tough material. They are much more fond of fungi and rotting organic matter, so they will be found in greatest numbers in a soil rich in organic matter or near the compost heap.

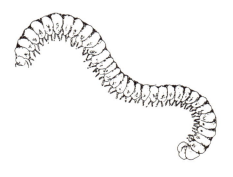

Snake millepede with two pairs of legs per segment.

Some of them are unable to digest the cellulose of plant cell walls so they have to devour large quantities of organic material to obtain sufficient food for energy. This means they begin to break down organic material which can then be attacked by other soil micro-organisms. Millepedes are protected by a cuticle which they produce using calcium and so they like soils with free calcium. They will suffer from desiccation though, and this is why in drought conditions, some of them are tempted to suck the moisture from plant roots.

Centipedes

Centipedes are distinguished from millepedes by possessing only one pair of legs per body segment and the number of segments varies between species from as few as fifteen to over one hundred. They have a tough outer coat but do not possess the water-proofing cuticle present in millepedes. They can move very efficiently with their many legs, and if you look at them closely you will see that they have made life easier by having each leg slightly longer than the one on the previous segment.

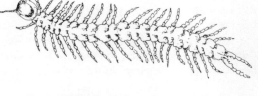

Centipede, with one pair of legs per segment.

Centipedes tend to be predaceous so their mouths are equipped with powerful jaws for biting, and poison claws which they use to immobilize their prey before eating. They feed on protozoa, mites, insects, slugs and worms, and they are also cannabalistic, especially if the other centipede is wounded. Centipede eggs are usually soft and laid in the soil so they are not easy to find. The young centipedes develop by a series of moults and in some species they look very similar to the adult except they have shorter legs. However, the young centipedes, which live under stones, begin life with only seven pairs of legs and it takes four moults before they have developed the normal fifteen pairs.

Symphyla

Symphyla are organisms very similar to millepedes and centipedes because they have an elongated body composed of fifteen segments, and twelve pairs of legs. The common species, found in greenhouses and gardens in Britain, is whitish in colour and about 8mm ($\frac{1}{3}$ inch) long. They are very active creatures, and tend to burrow deeper into the soil as soon as they are disturbed. They feed on dead and dying plant material, but also enjoy any plants with succulent root systems and completely destroy root hairs and small roots. These wounds often provide entry for bacteria and fungi, so plant deaths may be attributed to the wrong cause. Correct diagnosis of symphylid damage is best achieved by soaking the soil and plant in a bucket of water, as the symphylids will then float to the surface.

Symphylid, with twelve pairs of legs.

Woodlice

Woodlice are fairly distinctive creatures and cannot really be described as long and thin, but they can be confused with pill millepedes, so they have been included in this group. The remainder of the organisms which belong to the same class, the Crustacea, are water organisms, so woodlice tend to require moist situations.

They have oval-shaped bodies with the back protected by tough protective plates. Their lower surfaces show seven pairs of walking legs and two pairs of antennae, although one pair is very small. One common species is known as the pill-bug, as it rolls into a ball when it is disturbed.

Woodlouse, with seven pairs of legs.

The female woodlouse lays her eggs in early summer and carries them around in a pouch underneath her body. When the young hatch, they continue to be carried around by the female until the first moult. They are white when young, and only have six pairs of legs.

Woodlice mainly feed on rotting plant material but they also feed on rotting animal material. They live in environments where they can retain moisture, and this may mean that you will find them under rockery stones, under cloches or in the greenhouse. If you are a tidy gardener there is probably not much rotting material around, so they may be tempted to nibble at roots, and stems at ground level, and seedlings.

4
SNAILS AND SLUGS

Slugs and snails are distinctive animals mainly because of the way
they move on their thick, muscular foot along a mucus or slime trail.
Few gardeners would have a kind word for these creatures as it is hard
to find any good aspects (except the praise they give you when they
have eaten well off your vegetables!). My soil is heavy clay so I see
many of these creatures, and although I find them hard to love, I will
try to show how you at least can find an interest in their life history.

Snails
Snails can be really attractive creatures because their shells can show
many distinctive intricate patterns. The shell is made of a tough outer
layer and an inner layer of calcium carbonate and this is why snails
are more predominant on alkaline soils. If you lift up a snail to
examine it, the head and body will rapidly be withdrawn into the shell
and it will produce a mass of froth which is unattractive to many
would-be predators. If you put the snail down again and wait
patiently, it will soon try to see if the danger has passed and will push
out its head. The head has two pairs of tentacles, and the eyes are
found at the ends of the longer pair. The eyes are not very good,
allowing visibility up to 100mm (4 inches), but they are light
responsive, enabling the snail to prefer shade during the day, and dull
light at night.
   The same tentacles respond to smell and taste and some snails have
been shown to sense lettuces from a distance of 600mm (2 feet) or so.
Their bodies have a tough leathery skin which is made up of a coiled

hump and a muscular foot. There is a type of lung within the shell, and you can see it open and close if you watch the area between the head and the shell.

Garden snail, with head extended showing the two pairs of tentacles.

Feeding

The mouth is made of two jaws, the lower of which has a long rasping 'tongue' known as the *radula*. This is composed of bands of hooked teeth about one hundred and fifty long by just over one hundred wide. These teeth gradually wear away from the front end, but will be replaced from special tissues deeper inside the mouth. The radula is very effective at rubbing off the outer tissue of leaves, then the snail can feed on the inner tissue. Often bacteria and fungi cause the plant material to rot and many snails would rather eat dead or dying material if it is available. However they will certainly attack living plants and graze off seedlings.

Mating
They usually hibernate from late October onwards, not emerging
again until spring. They mate usually at night during late spring – a
process where a pair exchanges sperms since they are hermaphrodite.
There is an elaborate courtship ceremony when the pair remain in
contact with the muscular feet. After a while they send a 'love-dart' –
this is a hard ridged, pointed dart which shoots into the skin of the
partner snail. This is followed by exchange of sperm. The eggs are
small, white and gelatinous and are laid in the soil in collections of
about fifty. They are covered over and left, and about twenty-five days
later small snails will emerge.

Slugs
Slugs are pests in most gardens except the very dry and even then the
British climate usually provides a damp season at some time during
the warmer months. They may have beautiful colours and decoration,
but few people spend long looking at them closely. They are very
similar to snails but they have no shells to escape into. A few forms
retain a tiny external shell at their hind end of the body, but in the
majority the shell is reduced to a stronger area within the muscular
body.

Slug, showing the tough, muscular body.

Feeding
They feed with a radula on leaves, stems, bulbs, tubers, fungi, algae
and animal material. They have been found to feed on aphids and
small flies, so they do have a beneficial aspect. They require a moist
environment so you will find them hiding under leaves or stones
during the daytime. If you remove dying vegetation you will remove
some of their daytime hideouts, but you will also remove some of their
food sources, so they may be encouraged to visit the more healthy
plant material.

Their eggs are similar to snails eggs and will be found in groups in moist soil or amongst rotting vegetation as these habitats provide protection from frost and drought. During warm weather the eggs will hatch in a few weeks, but during the winter they may remain dormant until the following spring. Young slugs look like miniature versions of the adult, and their growth rate is very greatly influenced by the environment. The garden slug is not mature for two years, but field slugs may mature in five months.

5

SPIDERS AND MITES

Mites and spiders are distinguished from the insects because they have four pairs of legs. Mites are very tiny and have only one body section, whereas the spiders have two body parts, a fused head and thorax and the abdomen beyond a waist-like region. I will concentrate on the spiders first of all, and then progress to the smaller but no less significant mites.

Spiders

Spiders have been the heroes of much folk-lore and tradition and although many people find spiders rather worrying, few people kill them intentionally. This is a good thing as they are carnivorous and feed on all stages of many of our insect pests. They are not equipped with antennae, but they have a pair of palps at the front of their head, and these are used as sense organs for smelling and feeling. Male spiders have clubbed palps, and in their case these have an additional use for inserting the sperm into the female's body. The jaws are pincer-like and include a poison gland which is used to immobilize the prey. Just above the jaws are eight simple eyes, two of which are used for long distance viewing, two of which give a wide field of view and the others help the spider to stalk and catch its prey.

Not all spiders produce webs, but they will produce silk for other purposes such as the protective cocoon around their eggs. The abdomen contains special silk glands, and the silk squeezes out through tiny pores in the spinnarets which are found at the tip of the abdomen. When a web or cocoon is being produced, the spider often guides the silk into the correct formation by using its back legs.

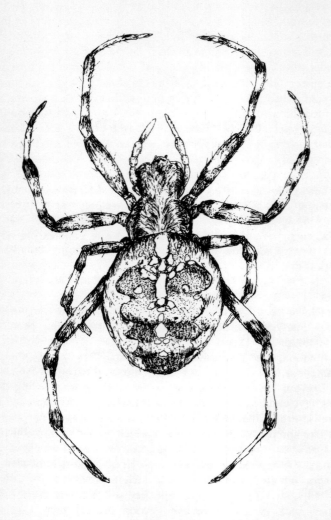

Spider, showing the two body parts, four pairs of legs and two palps on the head.

Web-spiders. Web-spiders are perhaps the most obvious especially on damp mornings when the webs have a silvery glow as each drop of water acts like a minute lens to reflect the light. The reason for these webs is to trap food, and at dusk you will often see the spider sitting in the centre of the web, and waiting. If some insect gets caught in the web it will struggle, and the movement on the fine silk strands warns the spider who will immediately trace the signal and find the prisoner.

The spider bites the insect and injects some poison to subdue its activity, and then she will wrap the captive up in a silk trap. She may eat the food immediately or some time later. You may have wondered why I have referred to the spider as 'she' – this is quite simply because the adult male spider spends his rather hazardous life courting rather than eating! He also has to approach the female spider via the web, and she frequently attacks him, so perseverance is essential. Male garden spiders usually manage to mate several times during the summer but by autumn time their energy is used up and they are often eaten by their last mistress.

Wolf Spiders. Wolf spiders are also common garden inhabitants, and they are mainly found scuttling in and out of low growing plants. They are called wolf spiders because their front pair of legs are longer than the other three pairs, and they use these legs to catch prey after a hunt. They like to rest in the sunshine, but they will always be on the look-out for any tasty morsel passing by and then they will stir into action. Wolf spiders care for their young in a more active way than the spiders which produce webs.

Once the female has laid her eggs and bound them into a silken cocoon, she picks it up with her jaws and carries it around suspended beneath her abdomen but held in place by a few extra strands of silk. The cocoon is often a very bulky addition to her body, but she will still wander in and out and over vegetation at great speed. When the eggs are ready to hatch in early July or so, the mother spider weaves a tent-shaped web and sits on guard outside. the spiderlings remain in the tent until moulting, and even then they may ride around on the mother's back.

Harvestmen. Harvestmen belong to the same group as spiders but their bodies are all in one piece and they tend to have very long legs, the second pair being the longest. These legs are very important sense organs, used for detecting the scent of valuable food materials. They have only two eyes which project above the head end of the body. They are carnivorous animals, being very partial to a wide range of the smaller fauna. Many of them also enjoy fungi and rotting vegetation, especially as a source of moisture.

Mites

Mites are tiny animals often only just visible to the human eye, but if they are examined with a lens they are found to have a rounded body and four pairs of legs. Many of the mites feed on living plant material and have mouthparts adapted for piercing plants and sucking out the sap. Only a few are really serious pests, but there are also several important predatory mites, and others which live in the soil breaking down organic matter and fungi.

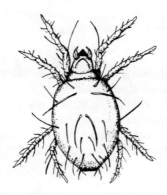

Mite: a tiny organism with four pairs of legs.

Mites as Pests

Red Spider Mites. Red spider mites are very tiny with bodies of 0.3-0.7mm (less than 1/50 inch) in length so they are noticed as tiny reddish dots moving around on the lower surface of leaves. There are various different species such as Bryobia mites, Fruit Tree Red Spider Mite and Glasshouse Red Spider Mite. All three attack quite a wide range of plants but the damage is relatively similar. The leaves show a fine yellow speckling and this may extend until the whole leaf is yellow or bronzed. They can also transmit viral diseases if they feed on an infected plant and then on a healthy plant.

The glasshouse spider mite produces a silk webbing which can stunt plant growth and cause distortion. The outdoor species survive the winter as bright red eggs protected from the winter frosts by being laid in cracks or crevices. When the young hatch the larval form has only three pairs of legs, but after the first moult they will be found to have four pairs. Maturity comes in about one month, so most summers there will be four or five generations.

Gall Mites. Galls are mentioned elsewhere in this book as abnormal plant growths due to some external stimulus, and many mites have this effect. One of the commonest garden galls is Blackcurrant Big Bud which is produced when the gall mites suck inside the buds and stimulate the bud tissue and contents to swell up.

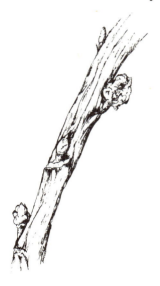

Mite damage produces Blackcurrant Big Bud.

Sometimes the bud may open to show abnormal leaves and flowers. If infected buds are left on the bush, the mites will develop fully, and they will spread to new plants. The mites can act as vectors for the viral disease known as 'Reversion'. Heavy pruning at an early stage will often prevent severe damage from Big Bud, but if virus infection ensues, the plants must be dug up and burnt.

Predatory Mites

A predator is an animal which eats other animals and usually predators feed on entirely different types of organism, but in the mites there are examples of 'cannibalism'. On fruit trees for example, where there are mite pests, there can be up to nine species of predatory mites. They seem to hibernate as adult females so only about ten per cent survive a normal winter.

In April-May they lay individual colourless ovoíd eggs on the lower surfaces of the newly expanding leaves. The eggs hatch to a larva with three pairs of legs, but after the first moult the nymphs have four pairs of legs. The nymphs and adults are predacious on spider mites and gall mites and their eggs. If there are no tasty mites available these predatory mites can survive by feeding on leaves, but they need animal protein if they are to reproduce.

The glasshouse red spider mite has been successfully controlled in many instances by using a predatory mite. The mites are yellowish-orange colour with pear-shaped bodies, and they develop at twice the rate of the pest. The female predators are very efficient at searching out adult red spider mite, and the young feed on the immature pests. The predator cannot live without its prey, so if it does the job very efficiently, it will eventually kill itself off, and new supplies of predator have to be introduced. As these are both mites, chemical control of the pest should be avoided because both the pest and predator will be killed by *acaricides* (chemical for killing mites).

6

INSECTS

The largest number of all organisms are found in the group known as the insects. Obviously there are many very different looking types of insects, but they have certain basic characteristics which unite them in the one group. They all belong to the *phylum Arthropoda* which describes the jointed legs and hard outer skeleton. The insect order is the largest one, but there are three other orders in the Arthropoda: the Arachnids which is the group containing the mites and spiders (see Chapter 5); the Crustacea, in which the woodlouse is included; and the Myriapoda which contains the millepedes and centipedes. These last three have all been dealt with in Chapter 3.

In the first Chapter I divided the common insects into different groups with a very brief description to help in identification. Only the adults show important insect characteristics (see diagram on page 17) and the many different larval forms may prove rather confusing. In fact you should find after a while that the larval forms also fit into general groups, and I suggest that you look carefully at the life cycle diagrams to notice the general features in each case.

APHID GROUP

Most gardeners are only too well aware of the aphids and it is unlikely that they are popular with anyone. They are obviously very successful as a group and they have been studied in detail to try to find some way of reducing their activities as pests.

Reproduction

Let us begin with their life history. They are far less complicated than many insects and undergo incomplete metamorphosis. This means that the aphid egg hatches into a nymph, or a small aphid similar to the adult except it has no wings and is unable to reproduce. That is the traditional way the aphid reproduces, but it is extra successful in that the female can also reproduce live young without mating.

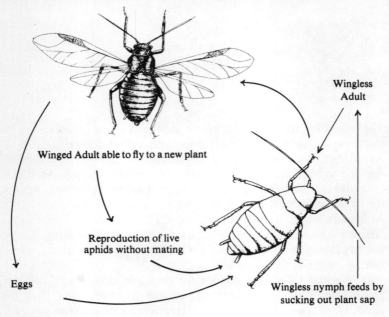

Winged Adult able to fly to a new plant

Wingless Adult

Reproduction of live aphids without mating

Eggs

Wingless nymph feeds by sucking out plant sap

Life Cycle of the Aphid.

This is why an aphid colony grows so rapidly, as reproduction is directly related to good conditions. Imagine an aphid having found your crop of broad beans – what an ideal place to bring up a colony! If you look closely at a colony of aphids you will also see that many of the aphids are wingless. This is another adaptation to an ideal environment – these aphids may well be mature and able to reproduce, but as there is a good supply of food there is no need to be able to move to another food supply. When the conditions start to get overcrowded in a colony, the new aphids will develop wings and can fly off to find a fresh source of food and start a new colony.

Feeding Habits

Aphids are specially equipped with mouthparts to enable them to suck the sugar out of plants (see diagram). They pierce through the plant cell walls with their stylet, and as the plant sap is under pressure they are able to absorb food without much effort. They tend to enjoy softer, more succulent plant material, and leave the tougher more woody plants to others. The aphids exude 'honeydew' from the anus — this is excess sugar from the plant sap because they have to feed on a lot of sap to get sufficient protein. The honeydew covers the leaves with a sticky layer and is often a food source for the black fungus sooty mould, which disfigures the leaves and reduces the light reaching the plant for food making. There are a lot of other important insects in a similar group to the aphids because their method of feeding is by sucking.

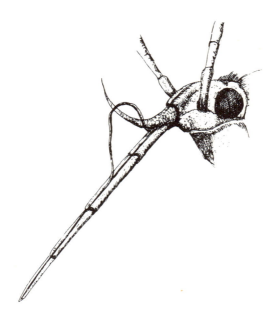

Sucking mouthparts found on insects such as aphids.

Froghoppers. These are the small nymphs which produce cuckoo-spit (the froth) on plant stems. The nymphs live in the froth to avoid drying out. The adults leap and the action is thought to look frog-like.

Leafhoppers. These are small insects found mainly on leaves. They are relatively host specific, which means each species only feeds on a small range of plants, but like the aphids they can produce honeydew encouraging sooty mould, and they are instrumental in the transmission of viral diseases. The nymph commonly leaves behind its skin at moulting, so they are often called 'ghost flies'.

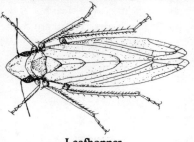

Leafhopper

Psyllids. These are commonly referred to as plant lice as they are very good at jumping because they have very long hind legs. They can be separated from leafhoppers as they characteristically hold their wings almost vertically when they are at rest. Most psyllids are specific to a particular tree or shrub, and they are often responsible for galls. Galls can be likened to cancer, in that a gall is an unusual growth of the plant cells stimulated by some other activity, in this case the sucking of the psyllid. The cabbage gall on box is produced by the activities of a psyllid.

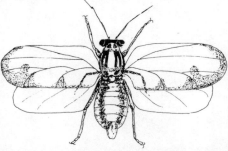

Psyllid, with long hind legs used for jumping.

Thrips. These are tiny insects which can be found in many different environments but one of the commonest is within flower heads. Within the flower, they pierce the plant cells with their stylet and suck out the plant juices. They may be found in very large numbers and can produce a mottled silvery appearance where they damage the cells, and they may transmit viral diseases. They can be beneficial by aiding in pollination, and on the garden scale are rarely considered as pests. They can easily be distinguished by use of a hand lens as their wings are fringed to form feathery structures, but you possibly already know them as 'thunder flies'. Some thrips are predacious on spider mites, so are certainly important for biological control (See Chapters 5 and 8).

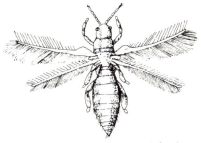

Thrip or 'Thunder fly' showing the feathery wings.

Scale Insects. These insects have a very descriptive name as they appear quite unlike normal insects. They are protected by a tough protective scale which is often very like the host material they are feeding on. They feed in a similar way to the other sucking pests, and can multiply very successfully by reproducing without mating. The scale protects the insect from predators and sprays.

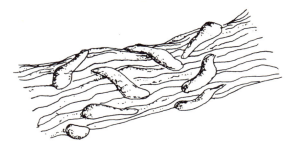

Scale insect merging into the stem structure.

Mealy Bugs. These insects are very similar to scale insects but they
are protected by a powdery white wax instead of a scale. They gain
extra protection by hiding within curled leaves, or sheltering at the leaf
base. Some mealy bugs live in the soil and feed on grass roots, but
they only become pests under glasshouse conditions.

Whiteflies. Whiteflies are small insects which are easily recognised
because the adults look like tiny white moths due to a powdery wax
coating on their wings. They are commonly found on the underside of
leaves, but they flutter as soon as they are disturbed. The female
whitefly lives about three weeks during which time she lays about two
hundred eggs. The eggs are laid in groups on the undersides of leaves
and each egg is in fact on a short stalk. The eggs hatch in about ten
days, and the nymph, which is flat and scale-like slowly crawls around
the leaf sucking out the plant juices as it moves.

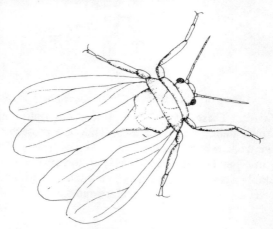

Adult whitefly which feeds on the underside of leaves.

As the nymph gets older, it develops a thicker waxy scale and
remains in one position on the leaf. Both the nymph and the adult are
plant pests mainly because they are the cause of secondary infection,
either in the transmission of viruses from one plant to another (see
page 105) but more commonly because the whitefly excrete a lot of
honeydew and this is followed by the growth of sooty mould (see page
57).

Capsid Bugs. These are included with the aphid group although bugs are generally larger and more complicated in structure, but they feed by sucking either on plant material (when they are classified as pests), or on animal material such as spider mites and leafhoppers, (when they are certainly beneficial). This immediately produces a problem if chemicals are going to be used to control pests which suck, as the beneficial capsids will also be destroyed.

Damaging Capsids. The apple capsid and common green capsid larvae are both pests as they feed on the young growth of apples, pears and soft fruit. This produces brown flecking and distortion on the leaves and corky areas on the fruits. Chemical control of the pest can only be used after flowering to avoid damage to pollinating insects, but needs to be carried out before the fruitlets are damaged by the insects.

Beneficial Capsids
Black-kneed capsids. Black-kneed capsids are fairly easy to recognize because they have a black band at the base of the long segment of each leg. The bugs are green in colour with reddish eyes. The adults

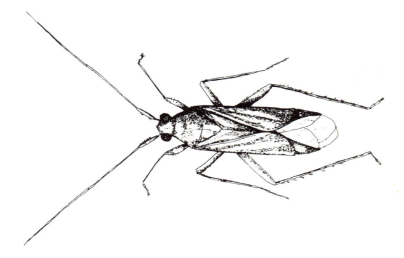

Black-kneed Capsid.

lay their eggs between July and October, the female slitting the soft tissue of apple wood and burying her eggs under the bark. The individual eggs (though she may lay fifty all told) hatch around late spring the following year, and the nymph spends the next month or so running actively over the leaf surface and sucking at food. Red spider mite females are particularly tasty, but they also feed on other mites, leafhoppers, aphids, thrips and caterpillars. There are many other capsids which are also found and they feed on mites, aphids and other small insects.

BUTTERFLY GROUP

Butterflies and moths are one of the easier insects to recognize, especially in the adult form when they usually have two pairs of membranous wings covered with tiny scales. The life cycle shows four distinct stages; the larva or caterpillar being in the feeding stage and possessing jaws for biting mainly at plant material. The adults feed on liquid food especially nectar, and so they are equipped with a long hollow sucking tube or proboscis, which is curled up under the insects head when not in use.

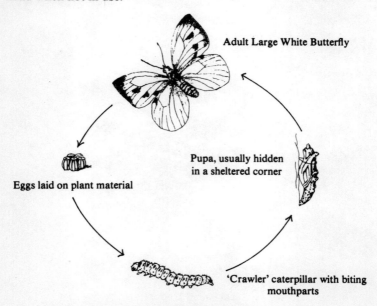

Adult Large White Butterfly

Pupa, usually hidden
in a sheltered corner

Eggs laid on plant material

'Crawler' caterpillar with biting
mouthparts

Life Cycle of the Large White Butterfly.

Butterflies and moths are commonly distinguished by their antennae. Butterflies in this part of the world, have knobs at the end of their antennae. Many people think that butterflies are good and moths are bad, but this is far from the truth. There are many fewer British butterflies than moths and the commonest ones are probably the white butterflies whose larvae are so fond of cabbages. Many of the moths do have harmful larvae, but there are many moths which we are rarely aware of as their larvae feed on weed plants and never cause us any harm.

Larval identification can be difficult, but a recognition of larval structure will help. Many caterpillars have three pairs of legs at the head end, and also have five pairs of prolegs or props which help them to crawl. The other caterpillars are known as 'loopers' as they have only two pairs of prolegs, and they move by looping the body upwards in an arch. (Do not confuse sawfly larvae with those of moths and butterflies see page 85).

A 'crawler' caterpillar, with five pairs of prolegs.

A 'looper' caterpillar, with only two pairs of prolegs.

Leaf Webbers. Some moths cause damage because the larvae form a protective web in which they feed. They can be found on various plants but are commonly seen on Hawthorn, Juniper, Cotoneaster, Prunus and roses. If webbing is seen early enough it is sufficient just to destroy the web to prevent further damage.

Leaf Tyers. The main offenders here are the Tortrix moths, the caterpillars which tie together leaves with silk, and feed away in safety. This means they are difficult to control, so it is best to get in first and destroy the affected parts to prevent spread.

Leaf Eaters. There are quite a large number of species which spend their larval stage feeding directly on plant material. They are less protected than some of the other species, but by the time they have burrowed into your best cabbage, they are difficult to get at. They can do a tremendous amount of damage to crops so you have to decide whether to use chemicals. If so be careful to avoid killing other beneficial insects.

A cabbage plant severely damaged by caterpillars.

Stem Borers. The larvae of some moths bore into the stems of many herbaceous plants such as Foxglove and Hollyhock. This may destroy the plants vascular system causing it to wilt and die. Others attack woody plants, but this only causes distortion and die-back. Once again control is best effected by removing the damaged material to prevent spread.

Leaf Miners. Leaf mines are made when larvae wander through leaf tissue separating leaf tissues and producing scars. Many of the leaf miners are moths although some are flies.

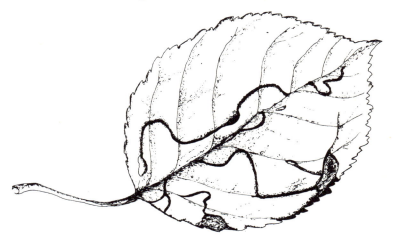

Mines on an apple leaf.

Apple and Rose Leaf Miner. The mines produced on apple and rose leaves are the result of the larva of a tiny moth — a closely related species producing the decorative mines on bramble leaves.

Privet and Lilac Leaf Miner. This is fairly common and the caterpillar produces the blister type of mine, but it emerges later and rolls the leaf backwards and binds it together with silk. These are found in June and a second generation in the Autumn.

Laburnum Leaf Miner. This larva tunnels into the leaf in May-June and mines a passage, but eventually settles down producing a blotch mine and causing the whole leaf to brown and die.

Cutworms. These are the caterpillars of various moths, but the larvae are all soft fat creatures which bite stems of plants at or just below ground level. This causes the plant to wilt, collapse and die, although the rest of the plant may be unharmed. The caterpillars are most active at night, and in the daytime they will be found hiding in the soil or amongst the vegetation. Most caterpillars of this group feed in July and pass through the whole life cycle rapidly but there are others who feed over the winter. The easiest way to control cutworms is to have a tidy garden, so there is nowhere for the caterpillars to hide during the day.

Swift Moths. Swift moth caterpillars spend two years feeding and they feed on roots causing wilting of plants. They will be brought to the surface by digging, and they are a popular food for birds.

FLIES

Flies are never very popular insects as they always bring to mind rather objectionable types like the blue-bottle, but as usual a little knowledge is a dangerous thing. Flies can be recognized from other insects by possessing only one pair of membranous wings, and the hind pair of wings are reduced to small knobbed organs known as halteres.

Next time you swot a blue-bottle, spend a moment or two looking at its beautiful wings, and the halteres behind. The wings are used for flight, and the halteres are important balancing organs enabling a fly to know when it has strayed from a straight route. Adult flies feed by sucking out juice from plant material or from decaying organic matter.

The larvae are very simple as they usually develop in a very sheltered existence, but many of them have mouthparts constructed of hooks which are used to tear the food material apart. (See Crane Fly life cycle on page 68). Fly larvae are legless and move around in a worm-like fashion, but they are basically very sedentary creatures more interested in feeding. When the larva is fully fed it pupates and fly pupae can be in various forms although many are encased in a tough protective puparium. Other pupae may show some of the features of the future adult, such as shadowing where the legs are developing, but often the pupal stage is tidily hidden away so you are unlikely to come across them in the normal garden routine.

Fly Leaf Miners. The larvae of some flies tunnel in between the tissues of leaves producing a line or blotch pattern on the leaves which can be quite attractive. When the larva wanders around during development, the tunnels are found to increase in width and although they begin as a silvery pattern, they turn brown as more tissue is damaged.

The commonest leaf miner is Chrysanthemum Leaf Miner because many people grow chrysanthemums in greenhouses and this enables five or six generations of leaf miner a year to get working. The mines are very small in the early stages which means they often escape unnoticed. Outside chrysanthemums rarely suffer any serious damage from this pest.

Holly leaf miner is perhaps more widespread. It can be recognised by yellowish-brown blotches on the foliage, but is fairly difficult to control by means of sprays due to the size of the plant and the waxy nature of the leaves. On small plants satisfactory control can be obtained by picking off the offending leaves and destroying them. The pupal stage is in the form of a puparium, which eventually hatches to the adult fly. Adults about 1.5mm (1/16 inch) long and are rather like black houseflies.

Root Fly. The adult fly is a dark grey colour and is similar to a house fly. She lays her eggs in late April or May, and when the larvae hatch out later, they move to the roots and feed for the next few weeks. They pupate in the soil, and a second generation of larvae may appear in late June. Plants attacked show poor growth, and when plants are small they may collapse and die. The commonest root flies are the carrot root fly, and the cabbage root fly which attacks most of the cabbage family but will also feed on Aubretia, Wallflower and stocks.

Adult carrot fly.   Damage caused by carrot fly larvae.

Crane Flies. Crane flies or daddy longlegs are common visitors to most gardens and become household visitors in the late summer, terrifying people by fluttering past them and getting trapped in hair or clothing. No doubt it is as unpleasant for them! There are several species of Crane Fly but only three species are real pests. The adults mate in late August or September just after they emerge from the pupa.

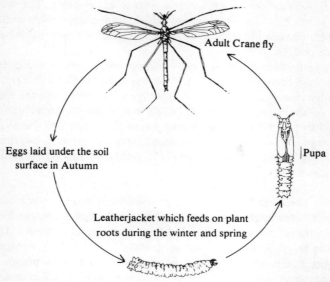

Adult Crane fly

Pupa

Eggs laid under the soil
surface in Autumn

Leatherjacket which feeds on plant
roots during the winter and spring

Life Cycle of the Crane fly.

The female lays her eggs in small groups just under the soil surface, especially where there is a good plant cover, such as a lawn area. The eggs will hatch a couple of weeks later and the larva or leatherjacket begins to feed on the plant roots. Leatherjackets can grow to about 35mm (1 3/8 inches) long and are legless with no distinct head and have a greyish-brown, tough, wrinkled skin. They require moisture, and many leatherjackets will be killed if it is very warm and dry soon after the larvae emerge. The larvae continue to feed throughout the winter and spring, pupating in the summer.

Leatherjackets can be controlled on a small scale by watering a lawn at night, covering it with a tarpaulin, and gathering the larvae the following morning. Alternatively insecticides can be used during mild, humid weather in autumn or spring.

Hover Flies. Hover flies are fairly easy to recognize in that they remain in one position for a while, and then move on to another spot by rapidly vibrating their membranous wings. They are commonly coloured with yellow and black markings. This is a defensive mechanism to confuse birds and other predators which mistake the flies for wasps.

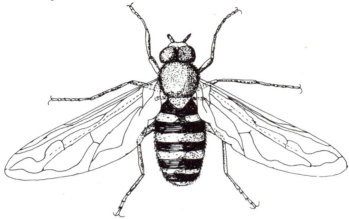

Hover fly, with distinct yellow and black markings on the abdomen.

Many hover fly larvae feed on aphids. Their eggs which are white or yellow are laid on leaves. The soft, legless larva emerge in a few days, and then uses the hooks around its mouth to seize aphid prey and then suck up the contents. The larva holds the aphid away from the plant so the prey cannot escape. The larva pupates when it has fed enough — this may be after it has consumed several hundred or so aphids — and it forms a hard protective case known as the puparium. In some species there may be several generations a year.

Hover flies appear to be directly affected by aphid colonies, and the females have been found to withhold egg laying while there are no aphids about. But they also lay many eggs when conditions are good, and they lay these nearest to the largest aphid colonies. The adult female fly needs to feed on pollen and nectar to ensure the ripening of her ovaries. The female is attracted to the aphids by scent, and is able to find aphids hidden within curled leaves or in a gall. She lays one egg at each aphid colony and the aphids hardly appear to notice her arrival, although adult aphids tend to record her with their antennae and walk away.

Aphids are often defended from predators by ants who appreciate the honeydew which aphids produce (see page 85). The common black ant will protect the black bean aphid, but the hover fly larvae counteract the attack by producing a protective slimy exudation. Hover flies are certainly very effective predators but their abundance is no doubt reduced by the use of chemicals and clearance of wilder areas.

Tachinid Flies. There are over two hundred and fifty species of Tachinid flies in the British Isles and the larvae of most of them feed internally on earthworms, snails, beetles, caterpillars and grasshoppers. Some of them are external parasites such as the blow flies of sheep.

Many of the female tachinid flies are able to incubate their eggs and larvae, so that the larva is active as soon as it is deposited on its host. The larvae hook on to the host and absorb their food directly, attacking the non-essential organs of the host. When the fly is ready to pupate it is then likely to kill the host. Other tachinid flies lay their eggs on host plants and the eggs do not hatch until they are eaten by the insect or larva.

Fungus Flies. Fungus flies get their name from their feeding habits as the larvae in particular feed on the fungal mycelium around rotting organic matter. The adult flies may feed on aphids and other soft insects. Some species have become pests in the greenhouse situation where the adults are attracted to organic fertilizers such as dried blood.

The larvae are about 5mm (1/5 inch) long with translucent bodies and black shiny heads and they feed on young roots of seedlings and cuttings. The complete life cycle takes four to seven weeks, so the numbers can build up quickly, but most gardeners will not often have to use chemical control methods.

Fruit flies are often found in similar situations, they are also tiny and tend to be found in clouds around a compost heap or, as suggested by their name, around over-ripe fruit. These flies are often used by schools and others for biology experiments on genetics, because they reproduce very rapidly. They are fairly distinctive with red eyes, but they are not harmful.

Gall Midges. Midges are tiny flies which often pass unnoticed in the normal garden. But the larvae of the flies which are small and maggot like feed within leaf tissue and this causes the plant to produce galls on the upper surface of the leaf.

The commonest ones found in Britain are the ones which affect Chrysanthemums in the greenhouse, the ones which attack violets, others which attack soft fruit such as currants, and many which attack the leaves of woody plants such as willows, ash and hawthorn.

The adult flies are fairly simple and often have orange-brown bodies and membranous wings with hardly any veins. The larvae are often orange also, but in most cases they do insufficient damage to require any form of control.

BEETLE GROUP

Beetles are the largest group of insects and there are many different species to identify; both beneficial and harmful. The front wings in beetles have become toughened to form the protective elytra or case, but they still have a pair of hind wings which they can use for flight. They seem to prefer to stay on firm ground, so they will often be found scuttling in and out of the vegetation. Little boys and entomologists alike, will bury jars in the ground (referred to professionally as pit-fall traps) and when the traps are inspected they will mainly contain beetles. The elytra protect the body and reduce water loss further, so beetles may be found in many different habitats and the garden is no exception.

Beetle eggs are fairly simple and usually pass unnoticed as they are hidden away near a food source or scattered randomly. They hatch out to the larval form which can be varied in type but always has a well-developed head and biting mouthparts similar to the adult. The cockchafer larva (see page 77) is very typically C-shaped with a brownish head, ferocious looking jaws and three pairs of legs which appear somewhat useless when you pick it up, but which enable it to move through the soil very easily. Another type of larva looks more complicated with the body showing the head, thorax and abdomen of a typical insect and they seem more able to look after themselves. Wireworms would come in this category (see page 76). The third type of beetle larvae is the soft type rather like caterpillars, hence the ladybird larva is a good example (see page 72). They can be distinguished from caterpillars as they do not have any false legs, just three pairs of legs on the thorax. The Bark Beetle larva is an example of the last type and is very simple, soft and unprotected as it is always found in a sheltered spot. After the larval stage, the pupal stage is more uniform. The pupae show distinct legs which can allow slight movement, even though they may often be found within a protective cocoon.

Ladybirds. Ladybirds are colourful beetles to look at and are one of the few creepy-crawlies which most people find quite attractive. They are also extremely beneficial as they are predators of aphids, scale insects, mealy bugs, thrips and mites. They are beneficial all through their lives, but many people do not recognize the larval and pupal forms. More than once I have been asked if a ladybird larva was in fact that of a Colorado Beetle!

Ladybirds spend the winter as adults finding shelter within plant material, under bark or within houses. In the spring they emerge from hibernation and search for suitable food material. The ladybirds mate

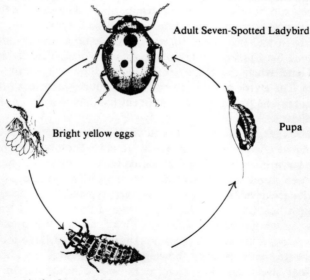

Adult Seven-Spotted Ladybird

Pupa

Bright yellow eggs

Active Larva which moults as it enlarges

Life Cycle of the Ladybird.

and the female lays bright yellow, elongated eggs in clusters, rows or individually; mainly on plant material, but sometimes on stones. The eggs hatch in two days to a week, and the larva emerges. The larvae feed actively on other organisms and they moult three times before hatching into the plump pupal stage. The pupal skin eventually splits and the adult beetle emerges.

Ladybirds have particular likes and dislikes and the Black Bean Aphid for example seems to be less nutritious and reduces the fertility of the ladybird. Large numbers of aphids are eaten by one ladybird larva. For instance a seven-spotted ladybird manages four hundred greenfly in five days. When a ladybird larva attacks one member of an aphid colony, the other aphids start to kick to try and push away the predator. Some may even attack the ladybird with a waxy secretion causing temporary paralysis. There are now some chemicals available which kill aphids but will not kill ladybirds, and if chemical control of aphids is necessary, these chemicals should be used.

Ground Beetles. There are over three hundred and fifty species of Ground Beetle in Britain, and they tend to be found in soil and organic matter. A few of them are pests, for example the Strawberry Ground Beetle, but the majority are important both because they breakdown organic material and are predacious on other pests. The larvae of ground beetles have well-armoured narrow bodies, and they remain in the soil or at the soil surface because they require damp conditions. The adults can be more active and the Violet ground beetle climbs trees in order to catch caterpillars of the Oak Tortrix Moth and Winter Moth.

Ground beetles have been shown to be very important predators of the eggs of cabbage root fly — one beetle managing an average daily meal of eighteen eggs! The beetle larvae move around with open jaws and feed on slow-moving soft animals such as earthworms and eelworms, and any animal debris.

Rove Beetles. There are over nine hundred species of Rove Beetles in Britain, and they feed on rotting plant and animal material, and some are predacious. They feed on root fly eggs and larvae, and the pupa stage of the root fly can be parasitised by rove beetles. Some rove beetles are predators of red spider mite, the larvae sucking out the fluid from the mites or sucking the eggs, whereas the adults eat the mites completely.

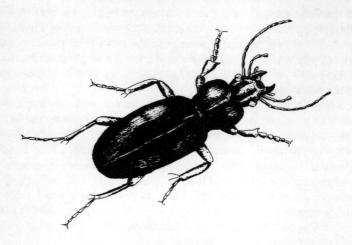

Ground Beetle.

Rove Beetle.

Bark Beetles. Since the horror of Dutch Elm Disease, bark beetles have become widely known. Dutch Elm is of course a fungal disease, but it is spread from tree to tree by beetles. Bark beetle adults and larvae tunnel between the bark and the living part of the tree. This destroys some of the plant tissue used for transporting food and nutrients around, so will tend to cause die-back and premature death. The female bores into the bark and lays her eggs in crevices in the tunnel walls. When the larvae hatch they burrow away from the original tunnel producing a typical radiating pattern. The larvae pupate in these new tunnels and emerge through the bark by tiny holes, and then fly off to a new site.

Simple beetle larva of the Bark Beetle.

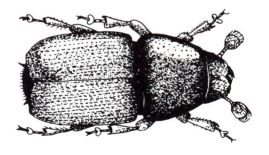

Adult Bark Beetle.

Weevils. Weevils are common visitors to the garden and are fairly easy to recognize as their head is extended to form a snout with the jaws at the end and the antennae coming out about half way along. There are over five hundred British species of Weevil and they all have these distinctive characteristics, but many of them have bodies covered with tiny scales and this produces attractive colouration. Adult weevils are plant pests as they bite off pieces of plant material. The larvae are legless, with curved white fleshy bodies and brown heads and tend to be found in a more sheltered environment. They are mainly found in the soil attacking roots, and quite often are found in seeds.

Adult Weevil.

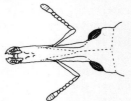

Adult Weevil head showing
the biting mouthparts.

Click Beetles. These beetles are more well known in the larval form which is the wireworm. Wireworms are important plant pests but they are not always as well known as they might be, and many an innocent centipede has been squashed in the name of a wireworm. Wireworms are shiny yellow or reddish-yellow creatures, with three pairs of legs on the thorax, and strong jaws on the head end. The eggs are found between May and July, usually in moist soil where there are plenty of fibrous roots. The larvae take about four weeks to hatch, and then spend the next four or five years feeding and growing.

Wireworm: the active larva of the Click Beetle.

They are most active in spring and autumn. Once they have fed enough, they bury deeper into the soil, usually in the summer, and pupate. Three or four weeks later they become adult, but they often hibernate until the following spring. The adult click beetles are brown, elongated in shape about 13mm ($\frac{1}{2}$ inch) long and they produce a characteristic clicking sound when they flick themselves over after accidently landing on their backs.

Chafer Beetles. Cockchafer beetles are also referred to as May bugs and are beetles about 25mm (1 inch) long with reddish brown wing cases. They spend the daylight hours resting on plant material and chew leaves and plant material. However it is the larvae which are more destructive. The eggs are laid in the soil, and when the larvae hatch five or six weeks later they start feeding on plant roots. The larvae hibernate during the winter and cause extensive damage the following summer. The larvae are fairly easy to recognize as the body is curved and the head is provided with extremely powerful jaws.

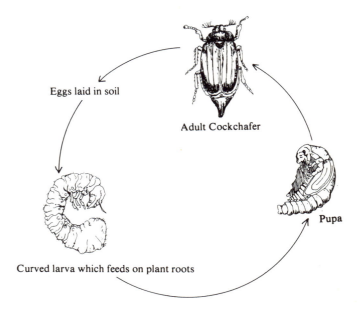

Eggs laid in soil

Adult Cockchafer

Pupa

Curved larva which feeds on plant roots

Life Cycle of a Cockchafer Beetle.

BEE GROUP

The main insects in this group are fairly easy to describe because the adult species have their distinct 'waist' between the thorax and abdomen, and many of them also demonstrate interesting social behaviour. The only difficult group are the sawflies because they do not have a waist, but as they have two pairs of membranous wings, it should be easy to separate them from flies and they are not like any other insects.

Perhaps you have already been surprised to find that many insects which you thought were pests, may in fact have a good side to them. Now we come to the bees, and I expect you feel quite convinced that the goodies are here at last. Well, of course, bees are marvellous creatures because they provide us with so much and are so important in pollination. But there are some bees which can be harmful to your garden, so we had better begin with them.

Leaf-Cutter Bees. Leaf-cutter bees look fairly similar to honey bees except that they are stouter and are covered in golden-brown hairs. Many shrubs are damaged by the female leaf-cutter bee, as she will cut out circular portions of leaf material and carry them away to construct the cells in which she brings up her young. The nest can be found in decaying wood or soil, and these bees are very independent and live alone. Each egg is provided with honey and pollen before the cell is sealed with the leaf material. The female is most active during June and July, and the new adults do not emerge from the pupae until the following spring. These solitary bees may be useful in pollination of plants with very open flowers especially in the Rosaceae.

The Honey Bee. Most people realize that we are indebted to the honey bee, not only for honey, but also because it is the timeless activity of these insects visiting flowers for pollen, that also fertilizes the flowers so the fruit will form. It is fascinating to watch them feeding, and they will only sting you as a defence mechanism not because they have anything against you! In fact their attack is suicidal, so it is best not to annoy bees for both their sakes and yours!

The bee colony is made up of three types of individuals, the queen, whose main job is to lay eggs, the drones which are the males who will mate with the next queen, and the workers who are sterile females carrying out all the social duties. Most of the bees which visit flowers will be workers, and they will collect the pollen on their hairy bodies. As they leave the flower, they brush their bodies clean with their legs, and then the back pair of legs collects all the pollen together. The worker then moves on to another flower of the same sort and continues until well laden. She returns to the hive where other workers will store it in cells.

A worker bee, showing the hairy body for collecting pollen.

When a foraging worker has found a particularly good source of food, she will return to the hive and do a dance which explains to the other bees how far away the food is, and in which direction to fly. Older workers also collect nectar, which is a dilute sugar solution produced by flowers. The bee sucks up the nectar with its straw-like tongue and mixes it with an enzyme which alters the sugar and reduces the water content. This mixture is taken to the hive where other workers change it into honey.

The queen bee is more matronly in shape than the workers, and she spends her time laying eggs. The eggs can develop into one of the two female types of bee depending solely upon their diet. New queen bees are fed on royal jelly; workers on mere honey. Drones develop from an unfertilized egg, and they are reared from Spring onwards, but at the end of the season they are not readmitted to the hive and so will starve to death.

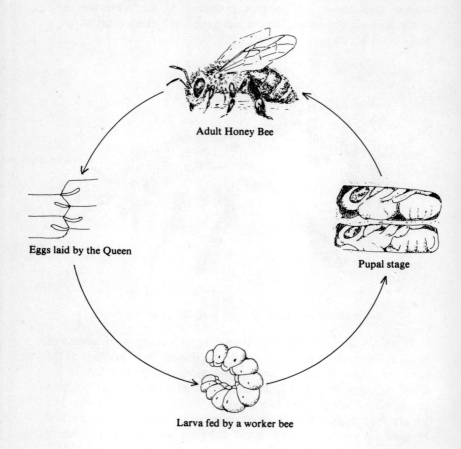

Life Cycle of the Honey Bee.

Bumble Bees. The life of a bumble bee is similar to the honey bee except that it makes its nests in holes or burrows under the ground. The mated queen bee is the only one to survive our winter, and when spring comes she comes out of hibernation and will look for somewhere to nest. She will make a wax cell and put a supply of pollen and nectar there before laying a few eggs. She will stay with the eggs, and when the young workers hatch as larvae, she will feed them so that they are fully grown within a couple of weeks. The workers take over food collection, while the queen busies herself laying more eggs. Both honey bees and bumble bees are essential for the pollination of many of our flowers.

True Wasps. True wasps are given that name to distinguish them from the digger wasps which will be dealt with in the next section. They can be identified by having crescent shaped eyes and they fold their wings lengthwise when they are not flying. The larvae are carnivorous, and feed on other insect larvae. The adult wasp likes sweet things, but they are not equipped with the nectar-sucking equipment of butterflies and bees, so they will feed on damaged fruit to get sugar.

Common wasps behave very like bumble bees although their nests are more elaborate as they are built up from thousands of paper cells. The nest may be underground, under the eaves, or anywhere similar where they are sheltered. The queen wasp scrapes off thin fragments of wood using her strong jaws. She macerates the wood and as it mixes with her saliva it produces a pulpy mixture. She makes a foundation, then returns to get more building materials. Once she has produced a small rounded structure with a few individual cells in it, she lays some eggs. She continues building until the eggs hatch into larvae, and then she has to find aphids and caterpillars to feed the larvae. When the workers emerge, they take over the job of nest building and feeding future larvae, while the queen spends her time laying more eggs. New queens will develop and mate before finding a dry corner to hibernate in, but all the other wasps die as the temperature begins to drop.

Digger Wasps. These are solitary wasps which keep their wings flat over their bodies when they are resting. They dig their nests in soil or rotten wood. They certainly should not be killed, as they feed on insects including aphids.

Ichneumon Flies. These are really incorrectly named as they are not flies, but another type of wasp. There are over one thousand eight hundred identified species, and the majority of the larvae are parasitic on the caterpillars of moths and butterflies. The adults have long antennae which are constantly in motion as the insect seeks for a likely food source. Female ichneumons often have a very obvious ovipositor, and it is with this that the parasitism begins.

Once the adult female finds a suitable host, using her antennae as a guide, she arches her body, pierces the host body with her ovipositor and lays her eggs inside. When the eggs hatch into the larval form, the larva feeds from the host's tissue, but initially this only reduces the host activity and does not kill it. When the larvae are fully grown they eventually kill the host, and this is closely followed by pupation.

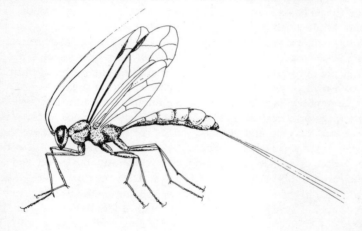

Ichneumon, showing long antennae and large ovipositor.

Braconids. Braconids are very similar insects to ichneumon flies, the main differences are found in the patterning of the veins of the wings. The particularly important one is a common parasite of cabbage white butterfly caterpillars. Each caterpillar may be host to over one hundred larvae, which kill the caterpillars just before they pupate. They produce bright yellow cocoons on the collapsed skin of the caterpillar. These should be left to allow the predator to complete its life cycle, and hopefully control future generations of caterpillars. Another braconid parasitises greenfly, and another the carrot fly.

Chalcid Wasps. These are common insects and there are several hundreds of different species found in Britain. They are usually less than 3mm ($\frac{1}{8}$ inch) long but they can be extremely beautiful with blue or green metallic sheens to their bodies. Most of them are parasitic on the young stages of many other insects, and they have been used successfully in biological control programmes (see p 112). The most important features for recognition are the elbowed antennae, and wings with very simple venation.

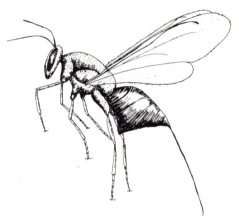

Chalcid wasp with elbowed antennae and simple veins.

One chalcid wasp, *Encarsia formosa*, has very successfully been used in the biological control of glasshouse whitefly. If you have a greenhouse or grow curly kale in the garden, you have probably come across whitefly (see page 60).

The scale stage of the whitefly makes it fairly safe from chemical attack, but the chalcid wasp lays her egg in a whitefly scale. The larva develops causing the scale to turn black, and the adult chalcid cuts its way out of the scale about one month later. The chalcid wasp mainly reproduces without mating, and if she cannot find a suitable scale to lay her egg in, she may 'sting' and kill older whitefly.

The wasp needs the whitefly to complete her life cycle, so if this is a method of control, it is necessary to have slightly more prey than predators, otherwise the predator will die out once all the whitefly have been controlled. However, the numbers of whitefly needed to satisfy the chalcid wasp, are not sufficient to cause severe damage to plants.

Cynipid Wasps. These are the wasps which produce many of the galls which are commonly found on oak trees. Oak Apple Galls are produced when the female cynipid wasp deposits her eggs at the base of a bud. This causes the bud to swell and by June or July it is fully developed. The adults emerge and after mating, eggs are laid in small roots in the soil, and another smaller gall produced. In the Spring about sixteen months later, wingless female wasps emerge, crawl up a tree and the cycle begins once more.

Spangle galls demonstrate another case where two different generations complete the life cycle. Inside the spangle galls the larvae feed and grow and eventually separate from their host in the Autumn. They continue to develop and pupate during the winter, female wasps emerging the following Spring. These lay their unfertilised eggs in male catkins or young leaves, and they produce small currant-like galls. In May-June the adults hatch and after mating they will produce spangle galls once more.

Robin's Pincushion is also produced by a cynipid wasp, and it is caused by the female laying her eggs in unopened rose buds. The plant produces a ball of moss-like tissue which is usually brightly coloured. Inside there are a mass of individual chambers where the larvae develop, remaining within the gall until the adults emerge the following spring and the cycle is repeated. None of these galls are particularly harmful to the plant, but they can be interesting to watch, and other insects may also make use of the galls as a protective home. Many cynipid wasps are also parasites of insect larvae.

ANTS

Ants show the typical feature of this group as their body has a definite 'waist'. Ants are usually wingless but at certain times of the year winged forms develop and a mating flight follows. Several species of ant are commonly found in gardens, making their nests in soil, under stones or paving or in old tree stumps. Most of you will have met ants when you have been in the garden, and you have been rudely attacked by a bite from their powerful jaws. Some of them sting, but many defend themselves by squirting formic acid at their enemies.

Ants are social insects living in a colony with a queen ant, male ants and worker ants. The mating flight usually takes place in August, and the male and female mate in flight. The fertilized queen lands and bites off her own wings, and then she digs a hole and lays a batch of eggs.

She feeds these ants and they are the first workers who will take on the job of looking after the colony. The queen ant spends the rest of her time laying eggs which are taken away by the workers and tended elsewhere. The larvae are fed on honey and insect grubs.

Ants are very fond of sweet materials and they are often found near aphid colonies as they feed on the honeydew produced by the aphids. Some species can be observed 'milking' the aphids, by stroking the aphid with their antennae which stimulates the production of honeydew. Sometimes they tend a certain colony of aphids, and shelter the aphids from fungi and other predators, and others are known to keep aphid eggs in their nest during the winter, and return the aphid nymphs to plant material in Spring.

Most householders think ants are dreadful organisms but in the garden, unless their nest is built just beneath a plant, they rarely cause much damage. They feed their larvae on other organisms and they aid in aeration of soil and mixing up different layers of soil. The wood ant collects much plant material to furnish its nest, and this is gradually converted to organic matter in the soil.

SAWFLIES

Sawflies belong to the same group as the ants, bees and wasps but they lack the distinctive waist. They get their name from the ovipositor because it is toothed like a saw and is used to cut slits in plant material before the female deposits her eggs. The wood wasp can penetrate wood with her sharp ovipositor. When the eggs hatch, the sawfly larvae which emerge are similar to caterpillars but they have stumpy prolegs all along their soft abdomens. They are pests at this stage as they feed on plant material.

Adult Sawfly.

Sawfly larva, distinguished by having prolegs along the posterior segments.

The commonest examples which are found in the garden are the Rose sawflies. The large Rose sawfly larvae are yellowish-green and covered with black spots, and in July they are found feeding on the leaves and biting out large areas from the edge of the leaf.

Another sawfly is the Leaf-Rolling Rose Sawfly, and this leaf-rolling is the result of the female cutting the leaf longitudinally with her ovipositor, and this causes the leaf to curl tightly around the egg. When the larvae emerge they feed on the leaf margin and work their way out. The slug sawflies are also found, and the larvae are very slug-like to look at and referred to as slugworms. The damage they do is distinctive, as once the larvae hatch they move to the upper surface of the leaf where they feed removing the upper layer of green tissue, but leaving behind the veins and lower tissue.

OTHER INSECTS

There are many other groups of insects which have not been mentioned in this book, but there are a few insects which must be included because they are frequent inhabitants of the garden. You should not find them hard to recognize, and they are worth getting to know.

Lacewings. Lacewings are distinguished from other insects by the presence of delicate membranous wings with an elaborate network of veins, and they also have very long antennae. Their bodies are usually brown or green in colour. They undergo complete metamorphosis, the egg hatching into an active larva with three pairs of slender legs, and curved hollow jaws which they use to catch other insects and suck out the body contents. The pupal stage is passed inside a silk cocoon.

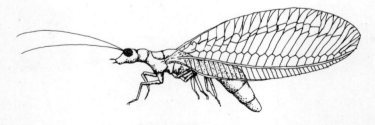

Adult lacewing.

Green Lacewings. These are often found in houses as they are attracted to the light. They have bright green bodies and green veined wings and a yellow metallic glow to their compound eyes. They fly slowly, and produce an unpleasant odour when they are handled. The green eggs are laid on stalks on a leaf surface. The larvae hatch and feed on aphids, leafhoppers and other small soft-bodied insects, and the lacewing larva may camouflage itself by covering its back with empty aphid skins. One larva is able to eat several hundred aphids during a fortnight, so they can easily influence an aphid population.

Brown Lacewings. These are smaller insects with brown or greyish wings. The larvae feed on aphids, thrips, and mites, but do not protect themselves with the skins of dead animals.

Powdery Lacewings. Powdery Lacewings are small fragile insects with wings less veined, the body and wings being covered with a white powdery substance. The adults look rather similar to whitefly, but when they are not flying they hold their wings in a steep arched position, not moth-like as whitefly do. The eggs are very tiny, salmon-pink in colour, laid on the lower surface of leaves or sheltered in crevices in the bark. After laying each egg, the female covers it with some of the white powdery material which covers her body. The plump larvae are almost colourless when they hatch but they have conspicuous red eyes.

They feed mainly on mites, and while they feed they attach themselves to the leaf by secreting a sticky substance from their abdomen. As with other lacewing larvae they pierce the prey with their hollow jaws and suck out the body contents. Adult female mites seem to be the most juicy, but in Autumn as the leaves begin to fall, the larvae are mainly feeding on mite eggs (which are the usual stage by which red spider mite survives the winter months).

Dragonflies. Dragonflies are one of the most attractive and easily recognised insects to be found. They have long slender bodies, large compound eyes and two pairs of delicate wings with a complex system of veins. They can have most beautiful colours, and as they are beneficial insects they need to be encouraged in every way possible.

The dragonfly nymph spends its life in water, so dragonflies are encouraged by ponds, rivers or nearby lakes. The adult dragonfly loves the sun and can fly well, so they can be found some distance

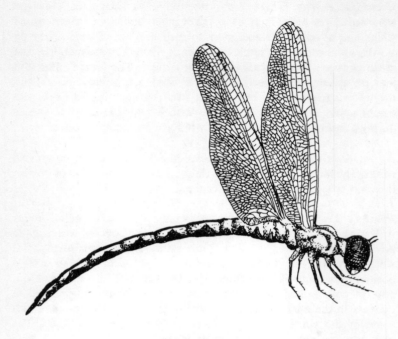

Dragonfly.

from water. The large eyes enable the adult dragonfly to catch its prey while on the move, and they are extremely partial to mosquitoes and flies as well as beetles and wasps.

Mating dragonflies are a fairly common sight if you sit near ponds and streams in Spring. The male searches for a female and when he finds her he holds her by the neck using his 'claspers' at the end of his abdomen. The female curves her body under the male until her abdomen is able to collect the sperm. The two may fly around together for quite some time. The eggs are dropped into water or laid in slits in the stems of water plants. The nymphs have long bodies, three pairs of legs and large compound eyes and a specially modified lower jaw. This is a long limb-like structure with two ferocious looking hooks on the end, but it has a hinge-like joint halfway along so the larva can hide the weapon close to its head. When it sees some suitable prey such as nymphs of other aquatic insects or even tadpoles or small fish, it shoots out this lower jaw, catching its prey and enjoying another meal.

Earwigs. Earwigs are well-known insects but there are only two species which are common in the garden. They have long bodies which are usually a shiny chestnut-brown colour, wings tightly folded so they are only seen on closer inspection, and two large claspers at the end of the abdomen. The name 'earwig' has led from a belief that as they like hunting out dark sheltered corners, so they may also hide within our ears and use their claspers to give the ear an unpleasant pinch! This does not seem to have much foundation, but maybe a reader has some evidence.

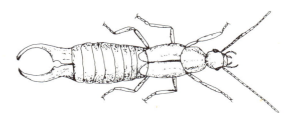

An earwig showing its large claspers.

Earwigs shelter in the soil as Winter approaches and they will mate there, then the female lays her eggs in an earthern cell. She looks after the eggs with loving care and will regularly lick them clean preventing attack by mildew. The young hatch in February or March, but remain with the mother until they are almost mature ten weeks later. There is commonly a second family later in the year.

Earwigs are omnivorous insects so they can feed on insect pests such as aphids, but their choice of food is closely related to the fact that they feed at night, and also that they need to be hemmed in by their surroundings. Chrysanthemum and Dahlia flowers satisfy this need, and if you grow these plants you have probably been saddened to find the flowers distorted or even prevented from opening. A sharp shake of open flowers will dislodge the pest and trapping can be a fairly successful method of control. It is usual to use an upturned pot on the end of a cane, and to fill the pot with loose straw — an ideal daytime retreat!

In most cases, gardeners are able to leave earwigs to their own devices and if no harm is apparent, it is more likely that they are being helpful.

Bush-Crickets. I was not going to mention bush crickets because I tend to think of them more as creatures of the countryside, but one was hopping around my garden the other day, so he obviously felt he should be included.

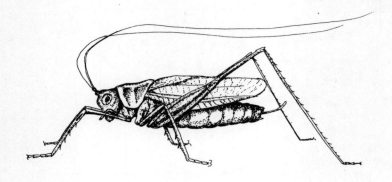

Bush Cricket.

Bush crickets can be distinguished from grasshoppers, because they have much longer antennae. They are attractive creatures to look at although they may surprise you when they jump, but they seem to find crawling involves less effort. They are beneficial because their diet includes pests such as aphids and other small insects, but as they hide during the daylight hours, you may rarely come across them. House crickets stray into sheds and greenhouses where they will happily feed on rubbish, but may have to spread to plant material if nothing else is available.

7

MICROSCOPIC ORGANISMS

Many of the insects and mites which have been described are very tiny and can only be studied properly using a lens or a microscope; but there are also many organisms which are even smaller. The majority of the microscopic organisms in the garden are beneficial, but there are some which cause our plants to suffer from diseases. Yet others can cause diseases in the pests, and this is one of the ways now being investigated as a method of pest control.

ALGAE

Algae are simple plants, of which seaweed is perhaps the best known. You may have some seaweed in your garden to forecast the weather, or you may use various seaweed extracts as fertilizers, but it is the microscopic algae we are going to consider here. The majority of algae in a normal garden will be present in the soil and puddles, but algae will also be found in ponds, on paths and on the damp sides of the tree trunks.

Soil algae are simple species which occur as single-celled individuals, simple filaments or colonies of simple organisms. Many of them are protected by a gummy material although the diatoms have elaborately sculptured silicate cell walls. Many of these algae are found in the upper soil surface where some light is able to penetrate, and they will photosynthesise building up sugars from carbon dioxide and water. Some algae live deeper in the soil and are able to feed on organic matter.

Algae flourish on damp sunny soils in Spring and Autumn, especially where the soil is fairly alkaline and fertile. They will add organic matter to a soil and help bind soil particles together. Some species also feed on atmospheric nitrogen to make their own protein, and this will improve soil fertility.

Algae in puddles and on tree trunks are not particularly harmful except if you find them unpleasant. They will produce a habitat for protozoa and bacteria. Pond algae are essential for aerating the pond to keep the water fresh and provide the fish with oxygen. Some algae multiply so rapidly that they can choke a pond and they must be removed, but they can, of course, be added to the compost heap.

BACTERIA

Bacteria are the types of organism which are often maligned as they are commonly associated with 'germs', but in fact it would be true to say that without bacteria the world would be a difficult place to live in. Our own digestive systems, for example, depend upon bacteria, and the gardener's world is a similar bacterial paradise. Yes, there are still some harmful bacteria which cause disease, but these are far less common than the beneficial variety.

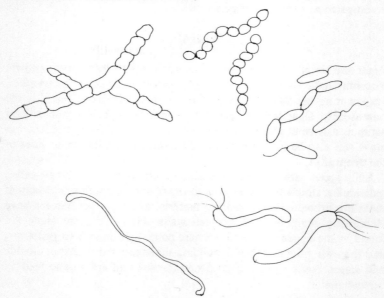

Typical bacteria.

Bacterial diseases in plants tend to cause cells to collapse and break down, which often leads to a bacterial slime. Bacteria have to enter the plant by some means so they are commonly the cause of secondary infection where the plant has already been wounded. If you take cuttings, the wounded stem is prone to fungal and bacterial attack, so you will find that rooting hormones often contain an additional antiseptic material. Probably the most important bacterial diseases are the bacterial wilts which include fireblight, and the bacterial galls.

Bacteria vary in shape but the three basic forms are the spherical ones or cocci which are about 1u (1u = 1/1000mm or 0.00004 inch) in diameter, the rods or bacilli which can be up to 10u in length and the spiral or spirilla which can be up to 50u long. They are simple unicellular organisms which reproduce by binary fission; one organism dividing into two when the environmental conditions are ideal.

Beneficial bacteria are very important in the normal garden and they tend to be found in two habitats, the soil and the compost heap. A single gram of soil may contain over one thousand million bacteria and these will all be active in maintaining a living soil. The commonest soil bacteria are rod shaped and many of them can swim by means of a flagella or whip. A lot of them are protected by a chemical sheath which may prevent the bacterium being eaten by protozoa, but also helps bind soil particles together.

Bacteria do not contain chlorophyll so they are unable to photosynthesize as plants do. However many of them are able to obtain their food from chemical sources, and these are then known as autotrophic bacteria. These bacteria are very important to the soil. The other bacteria feed on organic materials such as rotting leaves or animal material, and by this they not only release nutrients to the soil but they also improve the soil structure.

The Nitrogen Cycle
The nitrogen cycle is probably one of the best known cycles where bacteria are shown to be of great importance. There are however several common misconceptions about the nitrogen cycle, of which the most widespread is that peas, beans and clover are an essential part of the process. Leguminous plants are very important, but the nitrogen cycle occurs even when these plants are absent. So there is no need for you to turn your lawn into a clover ley!

Nitrogen is an essential element for the manufacture of protein in plants and animals, and animals obtain their nitrogen directly or indirectly from plants. Plants take up nitrogen from the soil in the form of nitrates, and they get the nitrates mainly by the activity of bacteria. If you want to give plants an extra boost, you may provide them with a nitrate fertilizer, but if it rains this will mainly find its way into the drainage system and encourage the water plants in canals and rivers. You may instead provide your plants with an ammonium fertilizer, or some form of compost which is broken down by saprophytic bacteria (those which feed on dead material) to produce ammonia. In this case other bacteria come into action and convert ammonia and nitrites into the important nitrates.

Nitrogen-fixing bacteria are the ones which feed on nitrogen gas in the soil air and change this into their own body protein. They are found free-living in a healthy soil, and then others are found living in the lumps or nodules found on the roots of leguminous plants. The nitrogen is not released as nitrates until the bacteria themselves are broken down by other bacteria. This is why it is important to leave your bean plants in the soil to allow the nitrogen to be released.

Unfortunately there are some detrimental bacteria which break down nitrates into toxic nitrites or back into nitrogen gas, but they are particularly prevalent in waterlogged soils, so it is best to encourage the beneficial bacteria by providing a well aerated soil, and then you should not be troubled by these denitrifying bacteria. If large supplies of unrotted organic matter are added to a soil, all micro-organisms will flourish which depletes the soil of air and the harmful bacteria will use up the nitrates.

The Carbon Cycle
Carbon is one of the basic constituents of living things and it enters the living world directly or indirectly by photosynthesis (food making) of plants or chemosynthesis of certain bacteria. This carbon is the main energy source for plants and animals, and our machines in the form of fossil fuels. If organisms die and rot down to become humus, bacteria are important both in the breakdown of organic material to produce humus, and then in the breakdown of the humus to release nutrients into the soil and return the carbon dioxide to the air.

Carbon to Nitrogen Ratio
Most gardeners build a compost heap to encourage the breakdown of organic matter into beneficial humus. If you are tempted to apply the

organic matter too soon, it will encourage harmful bacteria (see above) and will reduce the available nutrients in the soil. The reason is that the bacteria combine the carbon with nitrates to make new bacteria, and bacterial protein contains much more nitrogen than plant protein. So when the bacteria break up plant material, they will take nitrogen from the soil. They also use calcium to make new cell walls and this will make a soil more acid. When you make a compost heap, it is advisable to add some form of calcium or lime, and to wait until it is well-rotted before adding it to the soil.

The Sulphur Cycle

Sulphur is another essential nutrient for protein synthesis and if there is plenty of air in the soil, the organic sulphur compounds produced when organisms die will be acted upon by bacteria and changed, via hydrogen sulphide, into sulphates. Plants take up sulphates and so the cycle is completed. If the soil is waterlogged certain anaerobic bacteria will abound (they can breathe without oxygen) and hydrogen sulphide will tend to accumulate producing the typical 'bad egg' smell.

Bacterial 'Fertilizers'

Bacteria can be added to soils with fertilizers to increase the effectiveness of the nutrients provided, and release these nutrients to plants. This has been used in Russian farming for many years, and can increase plant yields. It is now possible to buy bacterial supplies to add to your compost heap, to hasten breakdown to humus which you can safely add to your soil. Alternatively, addition of living soil will add many natural bacteria, and they will soon flourish when surrounded by the wealth of food material.

Bacterial Diseases

Some bacteria are harmful because they use living plants as their food sources, and this means they digest living tissue to absorb the nutrients from it. The result is often a slimy mess. As bacteria are such tiny and simple organisms, they often enter plants which have already been wounded, so causing a secondary infection. They multiply most rapidly when conditions are warm and damp. Chemical control methods or bacterial diseases are seldom very satisfactory, so careful sanitation is essential to avoid infection wherever possible.

Bacterial diseases may cause disfigurement (such as bacterial leaf spots) but do not cause too much harm. But the damaging diseases are those which cause the plant to rot and wilt, often just around ground level where the environment is more conducive to bacterial growth. Fireblight is probably the most economically important disease as it is very damaging to apples and pears and it would be disasterous if it spread through our commercial orchards. This particular infection enters the plant through the blossoms, and then moves inside the tree to spread to other branches. Whole branches will die giving the impression of fire damage. It produces a reddish stain just below the bark, and in the early stages it may be satisfactory simply to prune off the infected material and burn it. The Fireblight Diseases Order 1958, requires any suspected case to be reported to the Ministry of Agriculture, Fisheries and Food.

Bacterial Cankers

Cankers can be produced by bacteria and one type is found on cherry and plum trees. It shows up as a sticky bacterial ooze on the bark in Spring, and leaf spots as a yellow ring surrounding brown tissue often followed by a 'shot-hole' (that is to say the dead tissue drops out). Branches may show dieback or buds may fail to break. When cankers are formed on herbaceous plants the leaves tend to discolour and wilt, and the stem becomes stained internally. The material collapses, but there is no evidence of mycelium or sporing structures as with a fungus.

Crown Gall on Forsythia.

Crown gall is a general name for a type of gall produced by bacteria on many different types of plant. The gall may be quite variable too, ranging from a massive distortion to quite tiny lumps. On herbaceous plants the gall or growth develops on the stem close to or just below soil level, but the galls can be found elsewhere on the plant. The bacteria enters the plant through wounds, and can remain in the soil while susceptible plants are present.

PROTOZOA

Protozoa are single-celled microscopic animals which need moisture to live in. They are present in the garden in ponds, puddles and water butts, but their most frequent habitat is the soil. Here they play an essential part in the general soil ecosystem especially because of their feeding habits. Many protozoa feed on bacteria and will control bacterial populations, and some feed on other protozoa. Soil protozoa are smaller than most water species as they are restricted to water held in the soil pores. Most of them can form cysts when conditions are particularly dry, but they can return to their normal state when there is sufficient moisture.

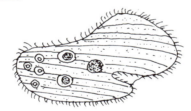

A protozoon which swims in soil moisture.

FUNGI

Fungi are usually described as plants although they do not contain green chlorophyll, so they have to obtain their food either by feeding parasitically on other living things or by feeding as saprophytes on dead material. The parasitic fungi tend to produce disease symptoms whereas the saprophytes are normally beneficial. In the garden evidence of fungi tends to be restricted to those which produce large toadstool-like fruiting bodies, or those which produce disease symptoms like mildew or black spot.

Beneficial Fungi

The most important fungi are found in the soil as saprophytes, feeding on organic matter and breaking it down to simpler substances. Fungi are more prevalent in soils rich in organic matter and they will be common occupants of the compost heap. They are tolerant of a wide range of pH, and will be more active in acid soils as bacteria will be less active and competition will be reduced. Many of the most simple fungi feed mainly on sugars and easily decomposable organic matter. They will often depend upon other organisms such as worms to break up the material and make it easier for the fungal mycelium (the feeding tissue) to enter. Soils contain many different fungal spores which are waiting for a suitable food source.

Fungi, feeding on organic matter.

Some of the larger fungi are able to feed on cellulose, a fibrous material, and lignin which is the basic constituent of wood. These fungi may grow quite slowly, but they are very important as they are the only organisms involved in lignin breakdown. It is not normally necessary to encourage fungi in the garden, there are always millions of tiny spores floating around in the air or water, and if they land upon your compost heap or some other organic matter they will soon set to work utilising the food source. But it is important to remember that fungicides are designed to kill fungi, and unwise use of soil sterilants may destroy some of the beneficial fungi, leaving you worse off than before.

Harmful Fungi

Harmful fungi are probably the ones you have come across and it is these ones which have given the remainder a bad name. Garden plants are susceptible to fungal diseases just as animals are, and some of these diseases can so cripple a plant that it seems kindest to destroy it. This is not just being 'soft' because while you have a diseased plant around, the trouble may spread to healthy plants. Fungal diseases often do not show until the fungus feeding tissue is spread throughout the plant, and by then control methods are costly and not always effective.

As I have done in previous chapters, I will reduce the fungal diseases to similar groups, and I will try to find examples which you will have come across.

Mildews. Let us begin with the mildews. There are two types of mildew, the powdery mildews which grow on the outside of the plant and send in special 'roots' to extract nutrients from the plant, and the downy mildews which grow 'down' inside the plant. In the early stages of infection, it is often possible to rub off the powdery mildew (they are never so damaging as downy mildews), but even so the upper surfaces of leaves may often be completely white. Powdery mildew is more common when plants are over-crowded and the soil is dry. Downy mildew is seen first as a yellowing of the leaf surface, but the lower surface shows a greyish mould which is really the reproductive spores.

Rusts. Rusts are another common group of fungal diseases, and they are aptly named as the infection shows up as bright orange or brown pustules. There are many different rust diseases, but on the whole they are host specific which means the infection will not spread to a different species. The main garden plants affected are Antirrhinum, Hollyhock, Pelargonium, Chrysanthemum, Mahonia, Leeks, Onions, Mint and Asparagus. If plants are particularly badly infected with rust, it is probably best to destroy the plant material. If possible, avoid these plants for the next couple of years as these fungi can only survive on living material.

Mahonia rust which causes the leaves to bronze.

Leaf Spots. The third types of diseases caused by fungi are leaf spots. These can vary from black spot on roses (a fungal infection which is common when the air is unpolluted — does that cheer you up if your roses are covered with black spot?) to leaf spots on rhododendrons, lupins, phlox and laurel. On some occasions, the infected plant material dies and drops out of the leaf and although this leaves behind a hole, it often reduces infection.

Cankers. Cankers are local diseases found most especially on woody plants where the infected tissue distorts, dries and splits open exposing the living plant tissue underneath. Cankers can be produced by bacteria, mechanical damage and woolly aphid, but the most common garden canker found on apple and pear trees, is the result of a fungal infection. The canker can be identified if it is looked at closely because when it is active there will be white pustules during Spring, Summer and Autumn, or bright red lumps over the Winter months. Both these structures produce spores which can spread the infection. Infection must enter through a wound, but although careful pruning is advised, leaf scars are open to infection for at least twenty-four hours from leaf fall.

Peach Leaf Curl. Peach leaf curl is a rather attractive looking fungal infection of peaches and almonds. The disease causes a plant reaction and the leaves become very distorted and bright red in colour. The leaves are likely to fall prematurely and the infection will reduce plant growth. If possible it is wise to remove the affected leaves before they get to the blistery stage which is when the spores are being formed. If a fungicidal spray is necessary, this should be done in early Spring before the buds begin to burst.

Honey Fungus. Honey fungus is one of our biggest problem fungi as it is a fungus which can live on dead and living tissue, and while feeding on some rotting material it can attack a healthy plant. The range of plants affected is huge, although some, such as Privet, are especially susceptible, but it seems to be most serious with trees and shrubs. It produces honey coloured toadstools to distribute its spores, but it also spreads through the soil many metres by tough black strings of tissue commonly referred to as 'boot laces'. Control methods are not very effective, so I hope your plants are never troubled by this disease.

Plant Wilting. The last group of fungal diseases I feel I ought to mention, are the ones which affect the transport system of the plant and produce wilting. Quite often these damage the plant at soil level where the soil is moist, but they are also inconspicuous until severe damage has been done. Plants often appear to be wilted during the day but recover in the cool of the evening, but later the lower leaves dry and eventually the plant dies.

There are also a set of diseases which attack seeds, seedlings and newly rooted cuttings, and they will also cause wilts at some stage in the proceedings as they interfere with water uptake. They flourish in the humid atmosphere required for propagation, and need to be watched for at frequent intervals so a control method can be carried out if necessary. It is often satisfactory with a small infection just to remove and destroy the infected material.

MYCORRHIZA

Mycorrhiza are the fungal associations found between fungi and the roots of higher plants. When these associations were first discovered, it was thought that such relationships were very limited to a few plants species, but recent work has shown that they are much more common than was first believed, and they have just been unnoticed. It is easy for us to concentrate on bad things and fail to acknowledge where beneficial organisms exist.

There are two basic types of mycorrhiza, the endotrophic mycorrhiza where the fungus grows amongst the cells of the root of the higher plant, and ectotrophic mycorrhiza, where the fungus is more like a sheath of tissue on the outside of the root. In either case the fungus soon becomes brown and muddy and may be difficult to distinguish from the root except by sectioning the tissue and using a microscope, although the roots colonised by ectrotrophic mycorrhiza tend to be very short and stubby. Heathers and orchids require mycorrhizal relationships to enable the plants to grow properly, but here the fungal tissue is internal.

This association is commonly referred to as *symbiotic*, in other words, both the fungus and the plant benefit from the relationship. The fungus needs the plant as a home and seems unable to live separated from the root system of a plant. The plants show much better growth where there is a mycorrhizal association, and this seems to be mainly because the fungus helps the plant absorb essential nutrients from the soil.

If fertilizers or manures are added to the soil, we expect to see better growth, but sometimes the results are not as dramatic as expected. This can be because the conditions are wrong for the bacteria which help in the chemical changes (see page 93), it could be because the soil is too acid or alkaline and the nutrients have become insoluble, or it could be that the plant roots are not very efficient at taking up nutrients.

One of the most essential nutrients for plant growth, is phosphorus, but it is also one of the most difficult to take up. Mycorrhiza seem to increase phosphorus uptake dramatically and although this process is not understood properly, it may be because the fungus increases the surface area for absorption. Such associations seem most beneficial when the soil is fairly low in nutrients.

VIRUSES

Viruses should really be considered in a separate chapter as they are defined as sub-microscopic since they are too small to be seen with an ordinary light microscope and can only be registered with an electron microscope. An electron microscope works by passing a beam of electrons through the material and if an electron hits something it will record it, so an impression of the object can be built up. Virus particles can be 'observed' by this method, but it is easy enough to see evidence of many viruses and their effects on plants.

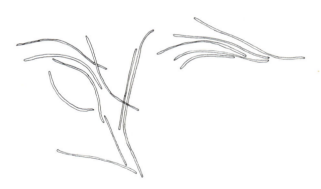

Virus particles.

The virus particle is like a part of a cell and some scientists believe that viruses are not living but simply a chemical. Living or not, they are able to multiply within living tissue and this usually produces disease symptoms. However, they are unable to move from one plant to another except with the help of a vector, and other organisms, especially insects, are important for this purpose.

Viral Symptoms

The typical symptoms which are produced often show up in the leaves first when this is where the virus enters the plant. Mosaics or mottles are very common symptoms of virus causing the chlorophyll or green colouring to break down exposing the yellow pigments instead. This may be very distinctive producing an attractive variegated leaf or a ring-spot, which is a ring of yellowed tissue. The other main effect of viruses is to cause stunting and distortion, because the virus uses up materials the plant would have used for normal growth. Viral symptoms may often be masked during the summer months as the better growing conditions tend to overshadow the inhibition by the virus. Some plants, especially many of the common weeds like groundsel, are known to have a latent viral infection, which means that the virus is present in the plant and can spread to other plants but has no apparent effect on the weed plant.

Virus symptoms on a rose leaf.

Virus Carriers

Virus transmission is dependent upon a carrier or vector, many of which are insects. Winged aphids are especially important here as they will pick up virus particles on the outside of their stylet when they suck at plants, and when they move on to another plant some of the virus particles will be carried over. Aphids are known as non-persistent vectors since virus transmission is simply due to a contaminated stylet, and if they do not feed on any other virus-infected material, they will soon be harmless again.

Another group of pests are more troublesome as once they have picked up a viral infection, it remains within the pest for the rest of its life. Again the trouble occurs with pests which suck plant sap, so the offending groups are the Thrips, Whitefly, Mealy Bugs, Leaf Hoppers and Mites. Nematodes are very important in transmission of viruses through the soil, and they can also harbour infection from one season to another.

The other main ways of virus transmission are where plants grow closely together and can become intertwined so that plant tissues become wounded and infected sap comes into contact with a healthy plant. The last main vector is man, and he can transmit plant viruses by touching diseased plants and then touching healthy plants, for example, when removing side shoots from tomatoes, or when taking cuttings from infected material.

Most gardeners do not need to identify individual plant viruses, but they need to be aware of virus types and symptoms. It is not practical to cure plant viruses, so if the infection gets a hold it is important to destroy the infected material and to try to prevent vectors from carrying the infection elsewhere. It is also worth discovering if the virus is liable to infect other plants as there are some viruses which are very host specific (which means they will only infect one plant species), and will not spread all round the garden.

Insect Viruses

Many butterflies, moths, flies, beetles and wasps are known to suffer from viral diseases during their larval stages. These have occurred naturally and effected a 'biological control' of what might otherwise have been severe pest damage. This has since been used as a method of insect control on lucerne, where the diseased caterpillar shrivels up but remains infective to its companions. Another virus has been found which affects cabbage caterpillars causing the larvae to burst and spread the infected body contents over the plant and soil. The uninfected caterpillars are also tempted to eat the dead remains of their brothers and sisters and this also encourages spread. This seems an attractive proposition as a method of control for the future, as this virus is specific to the pest and leaves no residues harmful to man.

8
CONTROL AND THE ECOSYSTEM

By the time you get to this chapter, assuming you have at least dipped into the previous seven chapters, you must be beginning to realize that our gardens are full of living organisms, many of which go completely unnoticed, and many of which are extremely useful. I hope you have been stimulated to look a little bit closer at some of the organisms and maybe you have also decided that it is wise to make use of the beneficial organisms. Some of the organisms are both beneficial and harmful when they feed both on our pests and on the helpful creatures, and at once a huge problem is upon us.

Ecologists have become almost a new breed of people as they have realized how important life is, and how the intereactions of living things, are vital to the well being of life. What exactly do I mean? Perhaps we need to look more closely at energy, this rather difficult concept, but upon which everything depends. The greatest energy source we know is our sun, and this directly or indirectly supplies our energy needs. Plants use the energy from sunlight to photosynthesize, or make sugar from the two available materials, carbon dioxide and water. This sugar is used by the plant as an energy source when it respires allowing it to grow and flower.

Animals are not nearly as efficient as plants, so they cannot utilize the sun's energy to make food, but they have to eat some plant source and use the plant's sugar for their food material. These animals are known as herbivores, as they are only one step away from the plant source. Other animals are carnivores, which means they feed on other

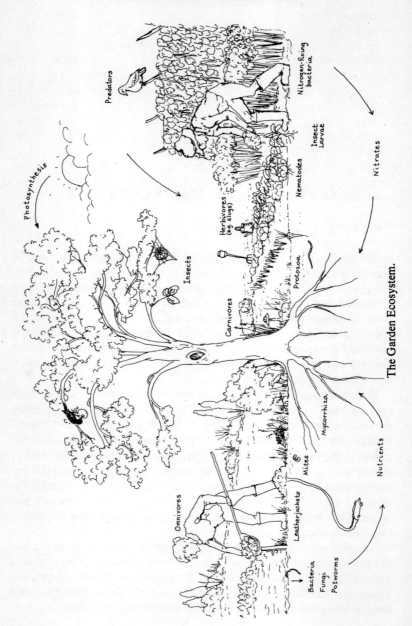

The Garden Ecosystem.

animals, so the food source will have been transferred at least twice by the time the animal uses it. Energy is lost each time the food moves from one source to another, so it is fairly easy to see that for a satisfactory existence there must be more plants than herbivores, and that carnivores will be in a minority.

This sounds fairly simple until you add on the fact that we are not the only organisms who are 'choosy' over our food. Several animals may be aiming for the same food, whereas others may be limited to one food source only, and if that is destroyed, they also will be destroyed. This inter-dependence of living things has been referred to as the ecosystem, and because of the large numbers of plant and animals species, it must be immensely complex. Even so the general principles can be shown in a diagrammatic form.

The Garden Ecosystem

If we now move to the garden, how does this fit in? First of all it is important to remember that gardens are unnatural places. You may or may not know what the land was used for before it became a garden, but if it had been left to its own devices it would certainly have been very different from its present use. Most people make at least some effort at gardening, so there will be a strange collection of plants all brought together. Weeds will often be pulled up or destroyed by weedkillers, yet other areas may be paved or made into lawns. All the plants will be producing food and all can be a food source for a herbivore (or an omnivore if you eat the fruit and vegetables!).

This motley collection of plants will encourage an equally mixed population of animals, some of which will flourish because their predators do not like these surroundings, but others will only eke out a meagre existence because some of their requirements are not fully met.

The ones which flourish and are, in your eyes, damaging may tempt you to try out some form of control method. Very few of our methods of control are specific to one pest, so beneficial organisms may be killed too. Even where the control is specific, the death of the pest will result in a reduction of food source for some other organism and upset the system. The result is that a garden habitat is not particularly stable because of the original plantings and removal of plants (by you) but the animals which may start to reach a balanced state, will easily be upset when you (or your neighbour) decide to attack your enemies. The problem is rather like a maze without an exit, as once the circuit has been entered, it is difficult to know what to do or how to stop.

Know Your Garden
One of the most important things is to use the brain with which you
have been blessed. Remember that you have the greatest influence
upon the garden ecosystem, and with a bit of thought, although you
will not be able to make it natural, you should be able to reach a
happy compromise. The first rule is – know your garden.

This means that you should be able to notice as soon as any plant
begins to look sick, if there is a plague of one particular type of insect,
or when you are not getting pleasing results from fruit and vegetables.
Early diagnosis is always helpful, and often you may exert your
controlling influence in the least harmful way. This knowledge should
become historical, and it may be, that it is best to avoid certain plants
in your particular area or soil type, or that an earlier or late sowing
may avoid epidemic times.

Within this knowledge, you need to observe or find out about two
other things, one is to identify the pest or disease so that you can find
out its likes and dislikes and, related to this, is investigating plants
which are less susceptible to the problems.

Preventative Control
Most of the descriptions in this book have made some reference to the
organism's life cycle, in other words, the descriptions have shown how
the organism begins life, what changes it passes through and how it
survives the winter. You may think that if an organism lives in our
climate, it must appreciate all the seasons, but many organisms will be
killed by frost and winter temperatures, so they need some method of
survival. If we have a mild winter, many of the organisms survive as
adults when they would normally have been killed, and these are the
years when pests build up to large numbers very early in the season.
This observation should immediately produce a warning signal –
preventative controls need to be considered where winter weather has
been insufficient to reduce the numbers of pests.

It is also important to recognize the pest in its other disguises,
larvae are often pests and normally look very different from the adult.
Eggs are nearly always tiny and well-hidden and they are rarely easy
to identify except where the life cycle occurs rapidly and many
generations are found together. Perhaps it is a good thing that eggs are
not often come across, because the question of good or bad can
remain unanswered. It is important to find out when the organisms
spread from one plant to another, both the seasonal time and also

factors such as climate. It may be possible to alter the climate on the local scale, such as removing the leaves of the lower part of the plant to allow air movement and reduce humidity, and it may be worth instigating a control method just before migration to new pastures. Natural predators can be extremely useful in preventing the build-up of infection to damaging proportions, but it is necessary to identify the predators, being careful to encourage them and not kill them by other means of control.

Biological Control

Biology is the study of living things, so biological control of pests and diseases is making use of biological knowledge to control the offenders. The term has been restricted to the introduction of predators, parasites and pathogens (diseases), although there are some more recent concepts which will be considered at the end of this section. Biological control has been praised by many, most especially because it may enable harmful chemicals to be avoided, and although chemicals have saved many lives, the horror of a few disasters stays heavily in the memory. There are other advantages put forward, some of which are open to challenge, but there are also some disadvantages.

ADVANTAGES	DISADVANTAGES
Control is selective.	No 'quick' response.
Predator will multiply.	Difficult to supervise.
No problem of pest developing resistance.	Results may vary with external conditions.
	Other less important insects may become pests.

Throughout this book, mention has been made of those animals which are parasitic or predatory, or those organisms which can cause disease. The fact that these occur naturally should make us more sensitive and less keen to use non-selective pesticides. Mention was also made of some of the biological control methods which are successfully being used in this country. The research work is slow and confined to the areas of greatest need which means the gardener's worries will only be dealt with as an aside to work for commercial horticulture and agriculture.

Many of our other garden 'aids' have come from the same sources, for example, selective weedkillers because hoeing is rather tedious on the field scale, and pelleted seeds were introduced so that the multi-seeded sugar beet could be planted as individual seeds avoiding the necessity for thinning. It is already possible for the amateur to get stock for biological control methods, but it is still easier and cheaper to buy chemicals.

New Ideas for Biological Control

Methods of control which make use of an existing biological system have much in their favour, so many research workers are concentrating their efforts in this direction. Some of the concepts almost verge on science fiction, and it will be interesting to see how many of our problems will be overcome in these ways.

Female insects attract male insects by the use of scents referred to as pheremones, which she releases from special glands. These chemicals are distinctive to the insect and if they are extracted and synthesized they can be used to draw the male insects to a trap. If the male insects are reduced in number, the total population will be reduced to a level where damage is insignificant. An alternative to a central trap, is to have the chemical everywhere so the males are unable to find the females. This is aptly named the 'confusion' technique!

In Chapters 1 and 6, mention was made of the changes and moults which insects undergo before they become adults. Each stage is controlled by a hormone, and it would be possible to use these hormones as a chemical spray to prevent the insect growing up.

In Chapter 7, reference was made to fungi, viruses and bacteria which attack insects, and some of these have already been successfully used to reduce pest numbers. Some of these techniques may seem ideal at present, but on many other occasions our successes have been dampened by the increase in numbers of some other previously insignificant pest.

Pest and disease resistant plants are an ideal answer, and while plant breeders seek for better yields and other attractive features, disease resistance is often top of the list. One way is to make the plant chemically unattractive to the pest or disease, but the alternative is to breed physical features which make the plant less susceptible. This might include such features as a very thick waxy cuticle on the leaf reducing attack by viruses, bacteria and fungi.

We are very used to the idea of inoculation of animals to enable resistance to be developed. This is successfully being done with tomato plants which can be inoculated with a mild strain of Tobacco Mosaic Virus and this prevents them being severely damaged by a more virulent strain.

Cultural Control

This is the name given to a logical idea — grow your plants as well as possible and your problems will be minimized. I expect your immediate answer is 'But I do', and yet your garden seems overwhelmed by unwelcome intruders. I will explain which factors need especial care, and perhaps when you see the reasoning, you will find that there are some areas where your vigilance has not been all it could be.

Land Preparation

The soil can be described as the basic requirement for a garden and it needs to be treated with respect if you want to achieve good results. Many tragedies are blamed upon the soil, but it would often be more realistic to accuse the soil management.

If you want to introduce new plants or annual plants such as vegetables then, if possible, the land should be completely cleared and dug well before the winter frosts. This will remove the weeds which can provide homes for pests to overwinter on, and also remove weeds which can be a source of viral infection. The digging will bring insect eggs and larvae nearer to the surface, and this can be a way of killing them when they are exposed to frost, or as they dry out in the winter sunshine. Your friendly robin and other predators will appreciate your efforts. Some pests will be buried so deeply that they can never escape through the depth of soil on top of them.

It is as well to remember that predators will suffer in the same way. The clods of soil will break down during the frosty weather, so that when you dig the soil in Spring you will find it breaks up and this will encourage your plants to grow well which is a good start to any healthy garden. Addition of organic matter in the form of well rotted compost or manure, will also encourage strong healthy growth, and replace many of the nutrients which would have been returned to the soil if the plants had died in situ, rather than making a brief appearance on the dining table! Some nutrients, especially nitrogen, encourage leafy, sappy growth which many of the sucking pests enjoy, and better results are achieved when a fertilizer high in potassium is used.

Soil pH (acidity) is important for healthy growth too, and it was pointed out in Chapter 7 that fungi are the micro-organisms which predominate in an acid soil. The fungus which causes club root to the cabbage family is worse when the soil has insufficient lime, and liming the soil can often be a successful control method. Waterlogged or badly drained soils are also prone to more problems than well drained aerated soils. and cultivation and addition of organic material and a healthy crop of plants, will all help to improve drainage and aeration. On the garden scale it is possible to add sand to areas where clay is very bad, and once improvement is started, it will usually continue on its own.

Tidiness

Many of our pests and diseases need moist, sheltered conditions for some stages of their life cycles, so they appreciate the gardener who leaves around heaps of garden rubbish, old pots, etc. The problem is not that simple though, because if the garden is too tidy, the insects which might have eaten the leaves which were already dying, may need to attack the healthy plants! But if the number of suitable hiding places are reduced, this will reduce pest numbers.

Hygiene

Hygiene is important at all times, but care at the early stages may prevent a problem later on. Seed boxes and flower pots should be scrubbed clean or soaked in a sterilising solution especially if any infection has been present. Seeds should be sown in healthy soil or compost. Cuttings should only be taken from healthy material otherwise this can be a very quick way of multiplying your problems.

Sowing and Planting Dates

Some problems can be lessened by careful adjustment of sowing and planting dates. Early sowing may enable the plant to be well established by the time the pest attacks it so the plant is well able to cope. Late sowings may miss the main outbreaks of the pest or disease, and often during the warm summer months, the plants will easily catch up with earlier sowings. Sometimes it is worth destroying plants earlier than usual, to prevent a build up of pests to spread to other areas.

Species Choice

Some seeds and plants will be more expensive because the plants are resistant to some of the many troubles we have discussed. If your garden has suffered trouble before, then it is certainly worth paying slightly more for resistant strains. Remember that many of the seeds are produced as a result of an artificial breeding programme, so if you decide to keep some seeds for next year, they are unlikely to possess the beneficial properties. Virus-free plants for soft fruits are especially rewarding, so resist the temptation to have plants from a friend.

Crop Rotation

The idea of rotating crops is a distant influence from agriculture, and you may feel it has no bearing upon you in your garden. The most important place is in the vegetable garden, where it is best to plant your crops in different areas of the vegetable patch in succeeding years. This will help the soil as different species have different nutrient requirements, and also their differing root structures will keep the soil in a better condition. However, more importantly, any soil-borne pests or diseases will be more prevalent when the same crop follows year after year in the same place.

Physical Control

Sometimes it is possible to afford some measure of pest and disease control by physical means. Perhaps the simplest method is hand-picking, where the gardener has to remove the damaged material and destroy it before it spreads. This can be very important, but obviously it is most efficient when an early diagnosis is possible. Traps can be very successful on the garden scale, rolls of sacking can provide day-time hideouts for weevils and slugs. Grease bands around trees prevent wingless moths such as the Winter Moth and March Moth, climbing up trees to lay their eggs.

Chemical Control

Chemical control of pests and diseases is a rather emotional subject, and while it is so easy to realize that traces of DDT can be found in wild animals far from the original source, we may forget that many peoples lives were saved by the same chemical. Most of this chapter has concentrated on alternatives to chemicals, but on some occasions, due to our types of planting, we may have to resort to chemicals. The names are very complicated (although the trade name is usually pronounceable) and some of the recommendations seem very finicky.

There are good reasons for this, mainly resulting from government legislation which is designed to keep us safe. If the instructions are followed carefully you should suffer no ill effects, but obviously it is wise to use the weakest chemicals whenever they will do a satisfactory job. The chemicals are described according to the agent they are designed to control, the way in which they act, the family of chemicals concerned, their individual identification and lastly a trade name.

I will not refer to trade products in this text, or even refer to many individual chemicals because new products are always being introduced and these tend to be safer and more selective. The following summary should explain the basic terminology:

Fungicides

These chemicals are designed to reduce fungal infection either by killing the fungus outright, or by preventing new growth, while the chemical is active. Most fungicides act on the surface of the plant so it is important to get a good spray cover.

Inorganic Fungicides

These are the older type of fungicides although many still are used very successfully, and they contain sulphur, copper or mercury. They are very useful for treating seeds to prevent mildew. The safest chemicals are those containing sulphur.

Organic Chemical

These are chemicals containing carbon and hydrogen. Any other constituents may vary.

Dithiocarbamates

These are organic fungicides which contain sulphur, and they interfere with the respiration of the fungus. e.g. Zineb

Systemic Fungicide

This is a fungicide which travels within the plant tissue, and is ideal for the control of many fungus diseases. Unfortunately chemicals tend to travel up the plant, so root diseases are not so easily controlled. e.g. Benomyl

Insecticides

These are chemicals for killing insects. They are often specific to a certain stage of insect development.

Stomach Poison
  This is used for pests which bite plant material, so the plant needs to be well covered by the chemical. e.g. Lead arsenate

Contact Poison
  The chemical only has to touch the insect for control.

Fumigants/Volatile Compounds
  These chemicals produce a vapour which kills the pest.

Ovicides
  These are materials for killing eggs, but it has been pointed out that eggs are often well hidden and difficult to reach.

Systemic Insecticide
  These chemicals pass through the plant tissue and are ideal for controlling pests which suck out plant sap, such as aphids.

Organochlorine Insecticides
  These are organic chemicals which contain chlorine. DDT is an organochlorine chemical, but it is no longer available to the amateur gardener. The commonest example is HCH, and its main advantage is that it is persistent, effecting insect control for some time. It is degraded to a less toxic material. Some insects have built up resistance to HCH.

Organophosphorus Insecticides
  These are organic chemicals which contain phosphorus. They are very toxic to insects and of variable toxicity to man, but they are fairly quickly degraded to harmless products. They act upon the insect's nervous system, but as different chemicals work in different ways it is possible to kill certain pests selectively. Some chemicals are more persistent than others, so great care should be taken to follow the instructions, especially if you are going to eat the plants you have treated. Many of these chemicals are systemic e.g. Dimethoate.

Vegetable Insecticides
  These are insecticides where the active ingredient is obtained from a natural plant source. The advantage of these chemicals is that no new chemicals are being added to the environment e.g. Derris. This is a contact and stomach poison slowly affecting the insect's respiration. It also kills mites. It quickly breaks down in the air and light resulting in harmless products.

Acaricides

These are chemicals to kill mites and spiders. Some systemic insecticides are also acaricides e.g. Dimethoate.

Nematicide

These are chemicals to kill eelworms and usually have fumigant properties.

Molluscicide

These are chemicals for killing slugs and snails, and they are normally combined with a bait and served up in an attractive pelleted form.

Soil Sterilant

This is a chemical which easily vapourises to penetrate the soil pores and control soil-borne pests, diseases or weeds.

Wetting Agent

A chemical added to a spray mixture to improve the spreading of the pesticide over the plant surface.

It cannot be stressed too strongly that chemicals should be selected with care and used with even greater care. Many of these chemicals should not be mixed with others. Dilution should be as accurate as possible, and chemicals should be kept well away from food and children. Selective chemicals should be used where possible so that predators are not destroyed.

Integrated Control

Integrated control is a glorious name for a very simple, logical idea based on the three methods of control just referred to — Cultural, Biological and Chemical. These ideas need to be considered together rather than as alternative methods of attack, and integrated control is just that. Grow your plants well, encourage predators and use chemicals only if necessary, and with caution.

Happy gardening!

APPENDIX

Organism	Page	Disadvantages	Advantages
Algae	91-2	May be unsightly on paths or damp sides of trees.	Help soil fertility.
Ants	84-5	Sometimes loosen plant root systems.	Adults and larvae aerate the soil, Larvae feed on other soil organisms.
Aphids	55-62	Suck out plant sap. Encourage fungi. May transmit virus diseases.	
Bacteria	92-7	Some cause diseases such as Fireblight and leaf spots.	Break down organic matter to beneficial humus. Soil bacteria make nutrients more available.
Badger	24	Produces subterranean burrows. May eat fruit and nuts.	Feeds on slugs, snails, insects and rabbits.
Bark Beetles	71-7	Tunnel in vascular tissue of plants. May carry fungal infections to plants.	
Blackbird	29		Eats worms and insects.

Organism	Page	Disadvantages	Advantages
Braconids	82		Parasitize caterpillars, aphids and flies.
Bullfinch	31	Pecks out buds and soft fruit.	
Bumble Bee	80		Pollinates flowers.
Bush Cricket	90		Feeds on aphids and small insects.
Butterfly	62-6	Larvae feeds on plant material.	Adults pollinate flowers.
Capsid Bug	61	A couple of species damage fruit trees.	Most bugs feed on soft insects and insect larvae.
Cat	23	Uses soft soil for toilet activities.	Scares off rodents and birds.
Centipede	42		Feeds on protozoa, mites, insects, slugs and worms.
Chafer Beetle	77	Adults attack aerial parts of plant. Larvae feed on roots.	
Chaffinch	32	Pecks at flower buds.	
Chalcid Wasp	83		Parasitizes insect larvae including whitefly.
Click Beetle	76	Larvae (wireworms) feed on roots.	
Crane Fly	68	Larvae (leather jackets) feed on roots.	
Cutworm	66	Moth larva feeds on roots.	
Cynipid Wasp	84	Produces galls though these are rarely harmful.	
Digger Wasp	81		Feeds on aphids and other insects.
Dog	23-4	May cause scorch with urine.	Chases off cats and rodents.

Organism	Page	Disadvantages	Advantages
Dragonfly	87		Predacious on other insects.
Earthworm	35-7		Aerates the soil. Mixes organic matter with soil.
Earwig	89	Distorts Dahlia and Chrysanthemum flowers.	Feeds on aphids.
Eeelworm (see Nematode)	37-9		
Fly Leaf Miner	65	Larvae burrow through leaf tissue leaving a scar.	
Fox	24	Their earthworks may cause damage.	Feeds on insects, birds and rodents.
Frog	28		Feeds on slugs, worms and insects.
Froghopper	58	Produces cuckoo-spit.	
Fruit Fly	70		Feeds on rotting fruit.
Fungi	97-102	May cause diseases such as mildews or rusts.	Breakdown organic matter especially when woody tissue is present.
Fungus Fly	70	Feeds on the young roots of seedlings and cuttings.	Feeds on fungus, organic matter and soft insects.
Gall Midge	71	The larvae produce galls on leaf tissue.	
Ground Beetle	73		Feeds on insect eggs, eelworms and soft larvae.
Harvestmen	51-2		Feed on organic matter, fungi and insects.
Hedgehog	26	Eats berries.	Eats slugs, worms, mice and frogs.
Heron	33	Steals goldfish from ponds.	

Organism	Page	Disadvantages	Advantages
Honey Bee	78-9		Pollinates flowers. Provides honey.
Hover Fly	69-70		Predatory on aphids.
Ichneumon Fly	82		Parasitizes the caterpillars of butterflies and moths.
Lacewing	86		Larvae are predacious on soft bodied insects such as aphids.
Ladybird	72-3		Feeds on aphids.
Leaf Cutter Bee	78	Cuts out areas of leaf for nest construction.	Aids pollination.
Leafhopper	58	Sucks out plant sap. Encourages fungi. May transmit virus diseases.	
Mealy Bug	60	As leafhopper.	
Mice	25	Eat seeds.	Eat insect larvae.
Millepede	41-2	Feeds on roots, seedlings, germinating seeds.	Breaks down organic matter.
Mite (see Spider Mite)	52		
Mole	24	Produces mole hills.	Feeds on slugs, snails, millepedes, insects.
Moth	62-6	Larvae feed on plant material.	Pollinates flowers.
Mycorrhiza	102-3		Help plants absorb nutrients from the soil.
Nematode	38	Some damage plant tissue and may also carry viral infection.	Feeds on nematodes, algae, protozoa, bacteria and organic matter.
Pigeon	32	Feeds on seeds and seedlings.	

Organism	Page	Disadvantages	Advantages
Potworm	37		Breaks down organic matter. Feeds on fungi, bacteria and nematodes.
Protozoa	97		Help balance the soil micro-organisms.
Psyllid	58	Sucks plant sap. Often produces galls.	
Rabbit	25	Grazes soft green shoots. Burrows may be unsightly.	
Robin	31		Feeds on insects and worms.
Root Fly	67	Larvae burrow into root tissue.	
Rove Beetle	73		Feeds on insects, mites and organic matter.
Sawfly	85-6	Females 'saw' plant material prior to laying eggs. Larvae feed on plant material.	
Scale Insect	59	Sucks out plant sap.	
Seagull	31		Feeds on insects.
Slug	47-8	Grazes seedlings and leafy material.	Helps break down organic matter.
Snail	45-7	Grazes seedlings and leafy material.	Helps break down organic matter.
Sparrow	32	Damages brightly coloured spring flowers.	Feeds on insects.
Spider Mite	52	Sucks plant material and may transmit viral diseases.	Some mites are predatory on other mites.
Squirrel	25	Feeds on nuts, bulbs, fruit.	Feeds on insects.
Starling	30	Feeds on berries.	Feeds on insects.
Symphyla	43	Feed on soft roots and root hairs.	

Organism	Page	Disadvantages	Advantages
Tachinid Fly	70		Parasitizes, snails, beetles, caterpillars and grasshoppers.
Thrip	59	May transmit viral diseases. Produces silvering on leaves.	Some are predatory on spider mites.
Thrush	30		Feeds on snails and insects.
Tit	31		Feeds on insects.
Toad	27		Feeds on snails, worms, woodlice, beetles and caterpillars.
True Wasp	81	Feeds on damaged fruit.	Feeds on insect larvae.
Virus	103-6	Reduces vigour of plants, sometimes resulting in death.	Insect viruses attack larval stages and easily spread throughout the colony.
Vole	25	Feeds on bark, bulbs and berries.	
Web Spider	51		Catches and eats aerial insects.
Weevil	76	Adults attack aerial parts of plants. Larvae feed on roots or seeds.	
Whitefly	60	Sucks out plant sap. Encourages fungi. May transmit viral diseases.	
Wolf Spider	51		Catches and eats ground-loving insects.
Woodlice	44	Feed on roots and seedlings when there is nothing else available.	Feed on organic matter.

FURTHER READING

Bristowe, W.S. (1958) *The World of Spiders*, Collins.

Chinery, M. (1973) *A Field Guide to the Insects of Britain and Northern Europe*, Collins.

Chinery, M. (1977) *The Natural History of the Garden*, Collins.

Dahl, M.H., Thygesan, T.B. (1973) *Garden Pests and Diseases*, Blandford Press.

Dahl, M.H., Toms A.M. (1976) *Pests and Diseases of Fruit and Vegetables*, Blandford Press.

Darlington, A. (1968) *Plant Galls*, Blandford Press.

Debach, P. (1974) *Biological Control by Natural Enemies*, Cambridge University Press.

Edwards, C.A., Lofty, J.R. (1972) *Biology of Earthworms*, Chapman and Hall.

HMSO (MAFF) (1969) *Beneficial Insects and Mites*, Bulletin 20.

HMSO (MAFF) (1974) *Pests of Ornamental Plants*, Bulletin 97.

MAFF *Approved Products for Farmers and Growers* (Published annually).

Owen, D. (1978) *Towns and Gardens*, Hodder and Stoughton.

Riley, N.D. (1971) *Insects in Colour*, Blandford Press.

INDEX